Ham Radio is Alive and Well

By Gary L. Drasch, K9DJT

Gary L. Drasch

Copyright © 2022 Gary L. Drasch
All rights reserved.

Copyright secured under the Pan-American Convention

No part of this work may be reproduced in any form or any means, photocopying, electronic or mechanical, except by permission in writing from either the publisher or author. All translation rights are reserved.

Cover Design and Artwork by:
Gary L. Drasch

Second Edition

ISBN-9798421392903

Trouten Publishing™
Port Washington, Wisconsin 53074

Dedication

To my grandfather, Hilbert Zimpelman, otherwise known as "Buck" to family and friends.

Gary L. Drasch

Introduction

Maybe you are one of those baby boomers who obtained your ham license and got on the air regularly, and then let the hobby slide off to the side because of other priorities. Maybe you were exposed to ham radio through scouting or a Boys Club. Or maybe you had an aunt, uncle, or neighbor who was an amateur radio operator. Maybe you used the *Military Auxiliary Radio* System (MARS) while in Vietnam or had a friend in your college radio club. You always had an interest in ham radio, but not enough at the time to learn the Morse code and study for an *FCC* (Federal Communications Commission) exam. All of our life experiences are different, and having said that, please allow me to share how mine evolved with the hobby.

> I was 12 years old, and most kids my age were into sports. Not me. I was into ham radio. I already had my Novice license and obtained my General class at age 13. I spent four years totally immersed in the hobby. At 16, I passed my driver's test, and like most young guys, I discovered girls. Not only had I discovered girls, I had a CAR! Between going to school, working, and having a social life, my operating hours deteriorated rapidly. From there life seemed to get in the way even more. Commuting to technical college, full-time work, and later getting married did not leave loads of time for "hamming it up." I did try to set up a *shack* (radio room) in our first home but my old HQ-170C receiver and Viking Valiant transmitter were not cutting it anymore. It appeared the ham radio community had moved into *single sideband* (SSB) transceivers (transmitter-receiver), which were cool, but I didn't have extra money for a new station, especially with our first baby on its way. Over time I sold

Gary L. Drasch

the *boat anchors* (vintage radios) and thought, "Well...someday I'll..."

I expect many licensed hams acquired their ticket at a young age, maybe not as young as I was, but in high school, college, or the military. At the same rate, many people out there had an interest in obtaining a license but never followed through. Somewhere along the line, they found their dream job or wife (not necessarily in that order) and could not find the time. The family came along with all its joys and responsibilities. The job required travel and/or relocation multiple times. Then before they knew it, it was all over. The kids grew up and were having kids of their own. They are retired or about to; maybe their spouse is or isn't. The thing is, they now have time to do whatever they want. Many might travel, move to a lake home, learn how to fly, scuba dive, or get back to ham radio, maybe all of them plus more. Then they start wondering, is ham radio even alive? Is anyone doing it anymore? After all, there is email, smartphones, texting, Skype, FaceTime, Zoom, and suchlike things being conjured up by some young engineer who doesn't look old enough to drive. Mention "ham radio" to strangers on the street, and if they are young they will say, "What's that?" If they are older, meaning 50 plus, they will ask, "are they still doing that?"

Having returned to this incredible hobby at the end of 2010, I have trouble believing I was one of those who asked (of a now very close friend, Lyle, WE9R), "Are they still doing that?" I used to drive about 50,000 miles a year while I was working and did remain active using an FM (frequency modulation) VHF-UHF (very high frequency-ultra high frequency) radio in my vehicle. One of the main reasons was Lyle. We would go fly-fishing together and kept in touch with the mobile and handheld radios while on the stream. At the time, that is what I thought was left of the hobby, FM VHF-UHF radios, and repeaters (a device that retransmits a signal). But the more I talked with Lyle, the more I discovered ham radio, as I remembered it, was very much alive and well. Lyle spoke of doing contests and DX-ing (looking for distant stations). He talked about working remote islands and countries, QSLing (exchange of confirming postcards), and so on. I recall saying, "Really?"

During this time, I was in the process of building a cabin in northern Wisconsin, or as Wisconsinites say, "Up North." With it being close to finished and being there all by myself one evening, I thought of how neat it would be to have an HF (high frequency) radio right then. I envisioned myself sitting at the desk, watching the haze burn off the bog in the morning while talking with another operator across the country. After returning home, I went out and bought my first HF transceiver, a used Yaesu FT-450AT. I was immediately re-hooked! Along came a completely new learning experience. I discovered the advent of the personal computer, Internet, and cell phones did not detract from the hobby but added a whole new dimension to it, making it even more fun than it had been in the past.

My hope is to reignite the passion of radio communications back into all those who thought the hobby withered away. I want to invite them to get back on the air and experience the same joy I rediscovered. Likewise, I want to encourage those who thought they should have gotten an amateur radio license back when, but didn't. And most of all, for all those people to have fun!

If you allowed your ticket to expire, or you do not have a license, read on. The Morse code is no longer a requirement, and there are many study guides and resources available in helping you check off, *"Become a Radio Amateur"* from your bucket list.

73, Gary
K9DJT

Gary L. Drasch

Contents

Dedication   *iii*
Introduction   *v*

Chapter 1	—They're still doing that...	1
Chapter 2	—The PC and Radio	7
Chapter 3	—DX-ing!	17
Chapter 4	—Propagation	37
Chapter 5	—Contesting	41
Chapter 6	—QSL-ing	51
Chapter 7	—Digital Modes	63
Chapter 8	—So Many Things to do	73
Chapter 9	—From Shack Heaters to Now	107
Chapter 10	—Antenna, antenna, antenna	119
Chapter 11	—Beyond Lids	133
Chapter 12	—Ham Radio Speak	139
Chapter 13	—Before Pressing that PTT	151

Acknowledgments   *155*
About the Author   *156*
Bibliography   *157*

Gary L. Drasch

x

Chapter 1

They're still doing that...

Yes, even with the dawn of the personal computer (PC), the Internet, email, texting, Skype, ZOOM, and smartphones, ham radio is alive and well. And not just voice communications but with Morse code too, better known as CW (Continuous Wave) to radio amateurs. Radio Teletype®, abbreviated as RTTY and pronounced "ritty," is much more of a common mode of operation than years past. Although SSB has predominately replaced AM (amplitude modulation) for voice communications, it is common to hear some AM-ers on the 75 and 160-meter bands (groups of frequencies), proudly transmitting with their vintage gear. I reflect on being a kid and drooling over the Collins S-Line gear which I could not afford. Many kids as I, have grown up (or should I say grown old?) and today have the means of not only owning some of that classic gear, but collecting it as well. That is another niche in the hobby. The America Radio Relay League (ARRL), our NRA (National Rifle Association) of ham radio, whose name came out of relaying messages from one station to another, continues to host the NTS (National Traffic System) to stand ready in the event of a national emergency. Hams are continuing to do all the things they did years ago, such as rag chewing, DX-ing, contesting, and QSLing; and having four new HF bands to do it in. Besides radio amateur satellites, hams are bouncing signals off the moon (earth-moon-earth, EME) and even off ionized trails of meteors. A multitude of FM and digital VHF-UHF repeaters are currently interconnected around the globe via the Internet. Today, a radio amateur can even have a QSO (conversation) with an astronaut on the International Space Station (ISS). Not to mention the operators and radio clubs who set up temporary stations at schools internationally, enabling science classes to speak with the astronauts. Radio amateurs are also incorporating the Arduino and Raspberry Pi into the hobby. You can say radio amateurs are the unfeigned leaders in technology, leading the way to communicate without the need for any commercial assistance.

So, why does anyone bother doing ham radio with the deluge of new technology flooding our lives on daily a basis? Have you ever given any thought to the game of Chess? I would have expected a game nearly 1500[1] years old to have vanished long before the introduction of Pac-Man, Donkey Kong, Geocaching, or Pokémon Go. In fact, people continue to play chess face-to-face and online using either their PCs or smartphones. Why does anyone persist in playing golf? (Seriously, why do they?) Of late, it has been reported people are starting to listen to vinyl records again, saying they sound better than CDs—I never saw that one coming. The point is; the same holds true for ham radio. People are still doing it. And it is not just a bunch of old farts. There are retirees returning to the hobby but a few young and middle age people are getting into it too. As of 2019, licensing in the United States has grown to 755,952[2] from 654,695[2] in 2007. Of those, approximately 51% are Technician class licensees (an entry-level license), 23% are General class, 20% are Extra class, 5% are Advanced class, and 1% Novice class (more about licensing later). I'm thinking, no big deal. Better than half are Technician class. Are these CB-ers (citizen band operators) who are generating this growth due to the elimination of the Morse code requirement? Or is it the "Doomsday Preppers" who have been budding everywhere? The real question for me has always been; has the "traditional" HF amateur radio hobby TRULY grown? Looking back at my 1962 *Callbook* (a directory of radio amateurs), I found there were approximately 225,000[3] amateurs in the United States and about 100,000 outside of the country during that time. I cannot find a breakdown of the classes of licensees during that period, but what I recall is the majority were *General* and *Extra* class. There had been the non-renewable *Novice* class license, which was valid for one year, and there were a few *Technician* class licensees, but do not believe there were many. After all, except for the Morse code test, the written exam for the Technician class and General class were identical. Because the privileges were so much greater for those holding a General class, the logical step was to bypass the Technician class and move directly into the General. In any case, the Technician class license was originally created by the FCC to encourage experimentation in the higher radio frequency bands, i.e., at and above 145 megacycles (yes, cycles, because we were not using hertz back then) to operate all forms of radio-controlled airplanes, boats, and cars. Typically, the Generals went straight to the HF bands.

To determine if the number of the so-called traditional HF operators in fact grew, I decided to do a little math and combined all the traditional class licenses in use today, i.e., Novice, General, Advanced, and Extra class, and came up with 371,443. This group accounts for 49% of the 2019 ham census. Then I subtracted the 225,000 hams of 1962, from the 371,443 we currently have and came up with an additional 146,443 hams, or a growth of 65%. And that's just the traditional HF licenses, which satisfied my question, it did grow! Now, if we bring the Technician Class licenses into the mix, there is a total growth of 236% since 1962. That's a lot of people interested in ham radio!

I believe, as of April 2021, there are three-million[4] radio amateurs internationally. I say believe because different countries use different systems to determine the number of licenses. Nevertheless, comparing that number to the 100,000 found in the 1962 *DX Callbook* makes it phenomenal.

The exciting thing about a large number of Technician class licensees is that they are first starting to experience a very tiny piece of this extraordinary hobby. It is an opportunity for the seasoned hams to encourage and mentor these Technicians into upgrading to the General and Extra class licenses. Doing so enables them to start enjoying the HF bands, or as I say, "The real ham bands." I'm going to go as far as to say, any Technician class who does not upgrade and get into HF really hasn't experienced true ham radio yet. Several members in my local radio club have gone all the way up the ladder. They have studied and obtained a higher-class license by accepting help offered by other members. They have moved beyond the VHF/UHF repeaters and began talking with people around the country and the world. That's ham radio! I have a couple of friends who are taking it a step further by learning the Morse code, even though it is no longer a requirement. Why? Part of it might be nostalgic, maybe it is because CW works where voice communications are unintelligible, or maybe it is nothing more than just plain fun!

The wealth of information relating to this wondrous hobby is unbelievable. Naturally, its accessibility is all due to the PC and Internet. It ranges from sites specializing in operating aids, DX-ing, DX-peditions (hams operating from exotic locations), contesting, specialized clubs, forums on particular brand radios, to forums on the various brands of boat-anchors, technical sites, and specific calculators, to buying and selling equipment—it is all there for the inquisitive ham.

The latest technology is allowing radio amateurs to use a PC to operate and control their transceiver. It should be of no surprise. PCs and the microprocessor are being interfaced with almost everything nowadays, even refrigerators! RTTY, and the new digital modes, operate by using a PC, a communications program, keyboard, and monitor. It transmits and receives its data via RF (radio frequency) between radio amateurs rather than the Internet. In the same manner, it has brought us a means of exchanging emails from a yacht, or some remote location. It does this using the same PC radio combination but communicates with automated stations having Internet conductivity. There's even an app for smartphones, which allows radio amateurs to communicate via the Internet to an FM repeater, and then via the repeaters RF to other hams using a mobile or handheld radio. All this, and I failed to mention the most remarkable part of ham radio—the ability to communicate globally, point-to-point, from one radio amateur's antenna to another. For me, that's the awesome part of the hobby, from my antenna to an antenna somewhere else, without any assistance from a satellite, cell phone tower, or repeater. This thrilled me as a kid and I'm equally as thrilled being a senior citizen. All that, plus no one is required to pay any service fee! And if we boil it all down, it still remains to be the tried and true form of radio communications. It is of the old school, wireless without any assistance of equipment in-between. If there would ever be a catastrophic failure of the Internet, as we know it today, ham radio communications will keep on working. Although new technologies have been integrated into the hobby, they are not required to communicate to the other side of the globe. You may have already heard or seen the phrase, "When all else fails, HAM RADIO!" It is true. Give thought to the Internet with all its phenomenal communications capabilities. It is vulnerable to failure because, at the end of the day, its conductivity is dependent on the use of our antiquated telephone wires. It is no different than using the old telegraph wires draped pole-to-pole alongside the railroad tracks in the late 1800s.[5] I cannot imagine a disastrous failure of the Internet. My wife for one would be devastated. Cripes, she gets upset if it runs a little slow or there is a hiccup with her electronic tablet or cell phone! Consider all the people who are so accustomed to being instantly connected to loved ones and all of a sudden, in the blink of an eye, they would no longer have that capability. This is one of the reasons radio amateurs are so important to public safety. Are you familiar with the Military Auxiliary Radio System (MARS)? MARS continues to be tested on an annual basis, where the army, air force, and naval bases communicate with radio amateurs around the country using a *split* mode of operation; they transmit on their military frequency and listen to the hams on ham frequencies. In turn, it is reversed; hams transmit on their own allocated frequencies and listen to the military frequencies. The technique continues to work unquestionably well. Hams who participate might be surprised to receive a nice thank you from the various

Capt. Trenton Selah, KF6BIE, 345th Airlift Squadron, Keesler Air Force Base, MS, Nov. 7, 2011. (Google for reuse photo)

military branches they worked (spoke with).

So here we are, new technology working with essentially old-school technology. What has taken place during the past 60 years is that radio amateurs embraced the personal computer, Internet, and cell phones. They did not allow themselves to be eliminated from wireless communications, but rather integrated the new with the old. They invited new technology in and started to experiment by melding it with their radio communications equipment. Ham radio has become more interesting and exciting than ever before because of it. I was absolutely amazed when I returned to the hobby as an active participant.

As you read on, you will discover the Internet and PC greatly enhanced the hobby rather than diminishing it. My personal computer is without question as much of my radio station as my microphone and key. Many pictures of ham stations today reveal the PC monitor and keyboard being in the center of the operating table, with the radio equipment located on either or both sides of it. It's a good thing!

Chapter 2
The PC and Radio

If you read the "Introduction" to this book, you are aware I thought traditional ham radio had died due to the advent of technology. Boy, I could not have been more wrong. As I pointed out earlier, the PC is an integral part of the ham radio station today. Even the cell phone is used to assist us in operating and finding DX.

One of the apparent things in the shack compared to the 1970s and earlier, is that we are now able to connect a PC directly to later-model radios via CAT6 control. CAT is an acronym for Computer-Aided Transceiver (not a PC network type cabling). This standard was originally developed by Yaesu to interface a PC via its serial port. Finding a male DB-9 connector, or a USB Type-B port, on the rear panel of a late-model radio transceiver is pretty much the norm nowadays. You might ask, "Why would a radio amateur want to connect a PC to his radio?" Five reasons come to mind: logging of QSOs, selecting commonly used frequencies, moving to a cluster spot (explained later), operating digital modes, and contesting. The first one, a logging program, was the main reason I originally interfaced my PC with the radio. It is true, we are no longer required to keep a logbook, but even so, it's nice to have a record of all our QSOs. A logbook is especially useful in collecting wallpaper, i.e., awards (or as my wife says, "Another piece of paper suitable for framing"). The cool thing with these logging programs is having software buttons that easily allow us to date and time-stamp the start and end of a QSO. The program automatically collects the frequency, and mode of operation, right from the radio. Looking at a clock or the frequency of the radio is no longer required. The program I use also allows me to maintain a list of favorite (commonly used) frequencies. These might be for a group of nets (a gathering of operators) I periodically check into, or certain parts of the band I use regularly, such as the CW portion. For example, you might want to include the W1AW, ARRL, and code practice frequencies on each band in the list. Once set up, all you need to do is double-click the item of interest, and the PC switches the radio to the frequency and mode you stored under that selection. It eliminates numerous button pushes and dinking around with the radio. And maybe the best part of the software is the use of the cluster and the ability to double-click on a spot (a cluster posting) which switches the radio to the frequency and mode the DX station is using. If you are interested in operating using the digital modes, you will need a PC connected to the radio. This is an area where the PC made a huge impact within the hobby. Besides making it easier to operate RTTY, it brought us a bunch of fresh modes to use. Lastly, you'll need a PC interfaced with the radio to do contests.

An important feature of the logging software is the ability to track QSL cards and electronic confirmations. Each QSO has its own record in the database which contains the QSO data but furthermore, if and when a QSL card was sent or received. This also holds true for electronic verifications (we'll talk about those later). The program generates a chart using all this data to show how many countries (entities) an operator worked on phone (voice communications), CW, and digital. It then breaks it down even further to how many on each band, AND if they are confirmed by a QSL card or verified through the Log Book of the World (LoTW—an electronic confirmation). You will find even more charts, which show how you are progressing towards an award, e.g., what states you may yet need for the Worked All States (WAS) award, or what zones you might need for the Worked All Zones (WAZ) award. As we know, the PC makes it effortless to manage and track data, an attribute not available to ham operators years ago.

In addition, the logging program I use can record the audio of a QSO by clicking on a software button and saving it as a .wav file. This is accomplished by the sharing of the digital mode audio connection from the PC to the radio. It is a nice feature if you want a recorded copy of a memorable QSO or for those involved in handling emergency traffic (information). Some programs provide complete control of the radio, i.e., every button, knob, or dial on the radio, is emulated through the program. Each can be adjusted through buttons and slider controls displayed on the computer screen. Although my program has this capability, I do not use it. I remain to be one of those touchy-feely characters who still prefer pushing the buttons and turning the knobs. You will find many different logging programs in the market; some are freeware and others you need to purchase. You will even find some programs specifically for contesting.

Who would have ever thought of operating a ham station from a remote location? A ham did of course—go figure! A fellow member of the DX association I belong to, Don, K9AQ, runs his station, which is physically located in northern Wisconsin, from his home in the Milwaukee area via the Internet. The main reason he put his station at his cabin in the north woods, is that the area is extremely quiet (electrically) and he has more room for multiple antennas.

K9AQ remote station located in northern Wisconsin. Also used as his "cabin" shack when there. (K9AQ, D. Solberg photo)

Gary L. Drasch

He remotely controls the radio, amplifier, antenna switch, and rotor while monitoring the power output and SWR (Standing Wave Ratio) of the station via various software applications running on PCs at both ends. Recently, he set up his cell phone with an app that allows him to access and operate his north woods station no matter where he is. This is a perfect example of melding the old with the new. Operating a remote system similar to this is a blessing for those who live in a neighborhood with covenants. What about seniors who decided to move into an assisted living type of facility? Typically, such buildings are configured on multiple levels and have no means of accommodating any type of antenna system. Having a remote station allows a ham to operate from his new living quarters as if he was at his previous home, rotor and all. Ponder for a moment—none of this would be possible without the PC and new technology.

I am not going to delve into it here, but the Internet has been implemented on the VHF-UHF side of our hobby too. Many FM repeaters are linked together via the Internet using Voice over

The remote control "home" shack located in the Milwaukee area. (K9AQ, D. Solberg photo)

Internet Protocol (VoIP); programs suchlike *EchoLink* and *IRLP* permit voice communications inter-conductivity through a PC. These programs allow licensed hams to use their PC to access a repeater domestically, or for that matter, in any country. They can connect via voice from PC to PC if they wish, but other than accessing a repeater and having a QSO via RF, it strikes me as being quite far from conventional ham radio.

Let's step back to electronic logging for a moment. I will be discussing Field Day (FD) later, but for those of you who are familiar with FD, you might look back and picture how we operated years ago. One operator was calling CQ (calling all stations) on the radio, another person with a large piece of paper divided into columns was checking for duplicate call signs, and lastly, another ham was recording the contact in a paper logbook. It took three people to run one station. Well, this is another example of how the PC has positively influenced the hobby. Nowadays, clubs run multiple radios during Field Day having only one operator at each. A PC, running contest software, is assigned to each radio. As the operator makes contacts, he enters the call sign of the station calling him, along with the proper exchange (information), into the appropriate fields of the logging program. At this point, the program checks for a possible duplicate contact, and if it is not a "dupe," the operator, after completing the QSO, presses enter to log it. If it were in fact a duplicate, the program would have flagged him with a notice on the entry screen. The operator would then inform the contact and wipe the fields clean. In addition, all the PCs are networked together. This allows for viewing a real-time combined score for the club and shows who the lead operators are during the event. If I were operating a radio, I can see how many Qs (number of contacts) my friends are logging compared to myself. It's part of the contest which adds a little more fun to the event—along with a few bragging rights afterward!

Okay, if you participated in Field Day, or for that matter, in any contest, you may recall operators getting hoarse from calling CQ and replying to calls. Or maybe how tired his or her fist arm (hand-arm used sending Morse code) would get. Not anymore. In this case, it is not the PC that comes to the rescue, but rather the IC chips, which were evolving at the time, i.e., those having voice storage capabilities. A friend of mine, Gary, W9XT, developed a voice/CW keyer in 1993. "The Contest Card," as he dubbed it, can be found in his article, which was published in the September 1993,[7] issue of *QST* (ARRL magazine). Nine years later, in 2002, he designed a standalone version which he marketed, until just recently, through his company Unified Microsystems. It was the model VK64. Basically, it stored up to four CW, and four SSB, messages in its memory. The operator would prerecord the messages using his own fist or voice. Now, if he or she wanted to call CQ, she would simply press the appropriate message button; the transmitter was keyed and the recording played over the air. If she wanted to give an exchange, she again pressed the button assigned to it. Gary decided to discontinue the product because contesting software, and most of the newer radios, now have this voice and arm-saving device built into their products.

I cited the Callbook earlier. If you are already a ham, I am sure you knew what I was talking about. This invaluable ¾" thick magazine was the "Telephone Book" of radio amateurs. I managed to retain two copies, 1962 and 1976. The formal name was The Radio Amateur Callbook Magazine; it was started in 1928 and published quarterly in Chicago. There were two versions, the U.S.A., showing all "K" and "W" calls, and the DX edition, which was titled The Foreign Section, showing the call signs in other countries. I referred to it as a telephone book because it was laid out as such. The pages were divided into three columns, and all the call signs were listed in alphabetical order rather than by name. By locating the call sign, I could identify the name and address of each. It was somewhat of an operating handbook too. It had various tables in it such as Q-signals (a brevity code), phonetic alphabet, great circle maps, countries list, prefix-by-country list, time-conversion chart, and global standard times. And what would a telephone book be without advertisements? Indeed, it had those as well. If I had a QSO with someone and wanted his or her address to QSL, this is where I turned to find it. Because of technology, the *Callbook* has essentially been replaced with a website called QRZ.com, https://www.qrz.com/. Although the *Callbook* remains in business today, they only provide content in a digital format (CD-ROM).[9] Alas, it appears the *Callbook* has become another victim of the Internet. (One thing I never understood about the *Callbook* was that goofy Flying Horse on the cover!)

Fred Lloyd, AA7BQ, founded QRZ.COM in 1992.[10] Everything which could be found in the old *Callbook* can be found on the website. Not only is it an up-to-date database of the FCC amateur call signs, but each call sign record can be managed by the holder of that call sign. Licensees can write a biography, insert pictures, and comment on their achievements and interests. For example, if I were having a QSO, I can go to https://www.qrz.com/ and in seconds, look up the other person by his or her call sign. More likely than not, a wealth of information will pop up. The other operator might have pictures of his station, antennas, club memberships, boat, woodworking project, cabin, fish caught, and etc. Each record of the system is similar to a personal webpage. If you enjoy checking into nets and are interested in learning more about one of the other check-ins, you simply do a QRZ.com on him or her. Maybe you want to see a picture of the person or a map of where he or she lives. Without a doubt, there is a satellite view of the map, which sometimes shows the station's antennas. Maybe you're a DX-er (a ham who chases distant stations), and you want to find out if or how one of your contacts QSLs. Does he do paper QSLs? Does he use the Bureau (QSL routing system)? How about LoTW or eQSL (electronic forms of QSLs)? Does he use a QSL manager (a person who receives and sends cards on his behalf)? What address should you use to send your QSL? Also, the number of lookups on his record is a good indicator of how active or new this ham is. Has he had a few hundred lookups or has been it thousands? Earlier, I said, "the QRZ.com website replaced the *Callbook*, plus more!" The more is this. The tabs on the website are titled: News, Forums, Swapmeet, and Resources. Click the Forums tab and find Ham Radio discussions on Technical matters, Logging, Contesting, RV, and Mobile operations. Under the Resources tab, you will find Study Guides, Online License Renewals, License Wall Certificates, DX Country Atlas, Grid Mapper, and a Ham Radio Trivia Quiz. Items for Sale, Reports of Stolen Radios, Scams, and Rip-offs can be found under the Swapmeet tab.

Years ago, every shack had an assortment of maps hanging on its walls. The favorite was the ARRL Great Circle Map, a flattened globe of the earth. With the United States placed in the center, it allowed operators to find an azimuth heading for their beam antenna. In addition, virtually every operator had a map of the United States showing all the numbered call areas too. I have an old copy of the *Radio Amateurs World Atlas* in my shack which was also published by the *Callbook*. I no longer use it because of the computer. The PC has become so prevalent in our homes and shacks, that we presently have an entire array of mapping software available to us via the Internet or as an installed application. Alex Shovkoplyas, VE3NEA, of Afreet Software, Inc., created a neat DX Atlas program. When using the program, I can display a map as rectangular, azimuthally, or as a globe. Country call sign prefixes are prominently displayed and measurements are easily made using the mouse. I'm also able to place pins into locations I have worked. Entities may be sorted in three ways; by prefix, by country, or by island. Simply clicking one in the list eliminates the hunting and pecking for it on the map. It appears automatically! Both grey-line (explained later) and propagation are displayed in real-time by clicking a button. The map can be divided into CQ Zones (40 geographic areas), ITU Zones (75 geographic areas), Latitude/Longitude, and Grid Squares. It is available at http://www.dxatlas.com/DxAtlas/. Check it out sometime!

Although our radios, be a late model or boat-anchor, may operate without a personal computer, I find it difficult not to do so. I am at a point in operating where I even take a notebook computer with me when operating portable at my cabin—even though there's no Internet service. The main reason for the notebook is to do all my logging and enable me to operate the digital modes. Usually, people use a PC, electronic tablet, or smartphone to log their QSOs, even if it is not interfaced with the radio, e.g., using a boat anchor. In my case, or for any ham operating portable and not having an Internet connection, we have a choice of downloading and installing the QRZ.com database (there is a fee for it), or installing a copy of the Callbook database, on our PC. Either one allows us to do all our lookups as normal through the program.

Now, having said all that, I still consider the ability to operate the radio without a PC as being the backbone of ham radio and the key to its continued utility. Ham radio continues to be our country's number one public communications backup system. Paper and pencil worked in the past and will carry on into the future. It's just not as convenient, efficient, or fun.

Gary L. Drasch

Chapter 3

DX-ing!

I'm a DX-er—meaning, I'm the guy who is willing to sit for countless hours, tuning and listening for distant stations outside of the United States. Sometimes, I might spend more time watching and pouncing on a DX station spotted on the cluster. I might be checking different DX websites which provide information on when DX-peditions are planning to be active. I put up with the interference and bad behavior on a DX's frequency while I eagerly wade through the *pileup* (explained later) trying to understand the pattern of how the DX station is listening. I might belong to a couple of regional DX clubs whose main purpose is to help fund DX-peditions. I attend seminars and conventions all in the name of learning more on how to contact that elusive station on the other side of the globe. I will most likely have a vanity call sign, 1x2, or 2x1, to help quicken an exchange. I have multiple towers, beams, and an assortment of wire antennas strung all over, including some onto the neighbor's property. Needless to say, my station consists of at least one amplifier requiring a 240-volt outlet. More importantly, I have a transceiver that has two receivers and a bandscope to boot.

I spoke of the cluster earlier, as well as above, and described how the PC became an integral part of the station. I say that in part because of the cluster. Some operators view the cluster as a form of cheating. It can be interpreted in many ways, but I view it as an augmentation in looking for and operating DX. The cluster is a worldwide network of computer servers, where radio amateurs can post information about DX stations, or for that matter, any station they worked or heard. It is not a general bulletin board to post a bunch of junk on, but rather a place to share information of value to other hams interested in working the same station. A posting usually includes such info as the transmit frequency of the DX station, the mode they are using, if they are operating split, and how far up or down in frequency they are listening. The *spotter* is the radio amateur placing that information on the cluster with his call sign noted. Because a spotter can be located anywhere on the planet, knowing his call sign will give us an idea of where he is. For example, if a station spot is heard in India, it doesn't necessarily mean it will be heard in the United States. The DX call sign, frequency, and mode are most likely taken directly off the spotter's logging software. Or, she may have entered it via a keyboard through a telnet window. The spotter comment though, e.g., up 2, simplex, FT8, or PSK31, is typed in using either method and usually has a character limit.

I have found cluster apps available for smartphones that will alert me if a particular DX station I have been hunting has been spotted. My new favorite is *HamAlert* at https://hamalert.org/about. The developer is Manuel Kasper, HB9DQM. It will send alerts via SMS, Email, and Push notifications through the HamAlert app. One of its key attributes is the ability to set filters for DXCC, Call sign, IOTA, SOTA, WWFF, CQ zone, Continent, Band, Mode, Time, or days of the week. These filters can be set within the app or on the website using *triggers*. Here is my turn-on though—we can set HamAlert to continually poll our *Club Log*, or our logging program, as to what DX we are kicking the brush for. There is no need to specify! It knows what we need.

Two other favorites are *iDX*, developed by J. Marcio, PY4OG, and *DxSpot* written by Bob Chandler, WB2ETR. I've often allowed myself to be awakened in the middle of the night by an app alert to chase DX. Even if the alert feature isn't used, it is still great for checking the general band conditions and activity. You never know, it might be worth leaving the spouse at the mall while going back home to work some DX!

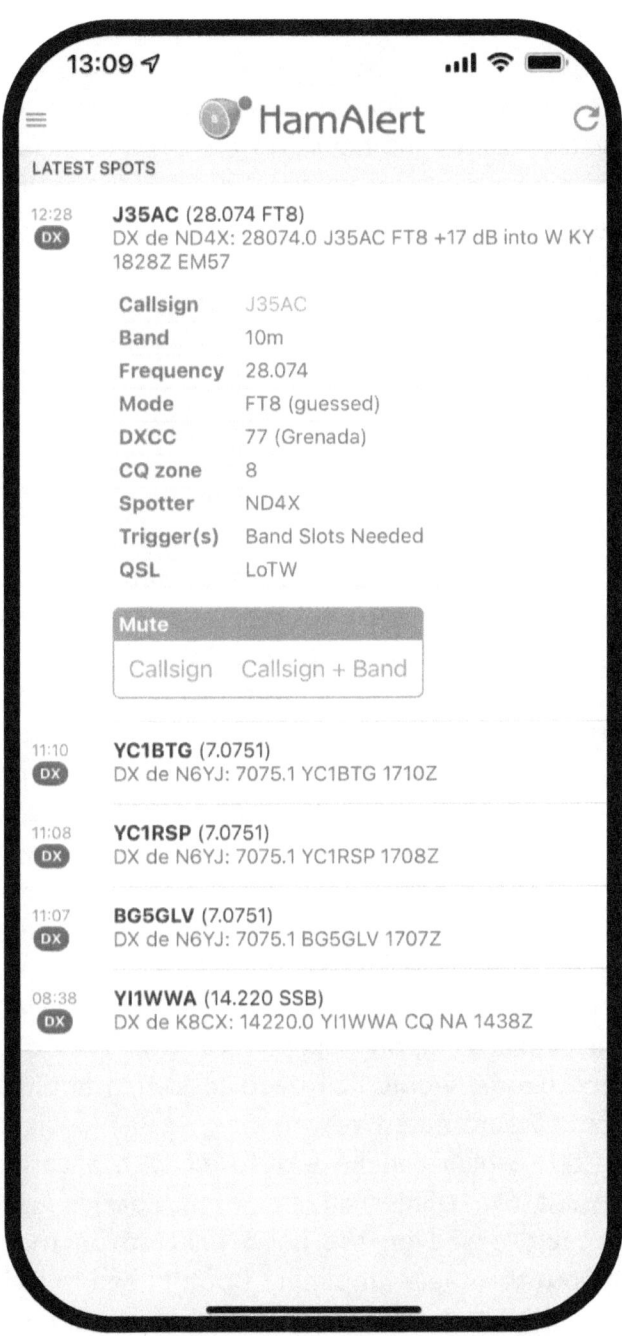

HamAlert screenshot on iPhone 13. The first spot is open which is showing the details. (G.

Any radio amateur may own a cluster node and make it available to other radio amateurs via the Internet. The differences between various nodes are usually the way the owner of the node decides who is allowed to use it, and how they are allowed to sort it. VE7CC has one of the best tools for sorting I've ever found. Check it out at http://www.bcdxc.org/ve7cc/.

Let's assume I am viewing the cluster within my logging software, and I see a spot for a DX station in a country I have not worked yet. How do I know I have not worked that country/entity yet? For myself, having worked 312 thus far, I guarantee it is not by memory. The logging software does it for me. It is constantly checking my logbook against the cluster spots, and flags me with an "X" and/or a red highlight on that particular spot. I get excited when I see those! Now I need to switch my radio to the mode and frequency spotted. The frequency is the frequency I will be listening to, plus I need to operate split because the spot said the DX was listening UP. For that reason, I need to adjust my transmitter VFO B to a higher frequency than he is transmitting on. It can take a moment or two, but this is the neat thing. I take my mouse and double-click the flagged spot on the cluster window and voila, the radio changes bands and switches to the correct frequency, changes mode, and splits the A-B VFOs. Wow! All that by simply clicking the mouse! The only thing left to do is adjust my "B" VFO to where the DX is listening and turn the beam. The software even makes that painless by displaying a heading to do so. Cooler yet, some software will even turn the beam for me when using an appropriate rotor control box. If my radio changed bands, wouldn't I need to switch antennas? Guess what, that can be automated as well. Gary, W9XT, designed a band-decoder in 2001 which can be interfaced with a variety of late-model radios. It can trigger an electronic coaxial switch to the appropriate cable based on the band selected on the radio. But wait, what about the linear amplifier? Don't I need to switch bands, and tune that too? Not to worry. Solid-state amplifiers, big ones, which switch bands with the transceiver and tune themselves are a reality. Even antenna tuners. So there we have it. Complete station automation. With a click of the mouse, we can change the band, frequency, mode of the transceiver, amplifier band, change antennas, and the direction of the antenna. This capability is not only highly desired by DX-ers, but contesters as well. Even so, we still need to speak into the mic or pound some brass (Morse code)! Or if we are contesting, maybe not. More on that later.

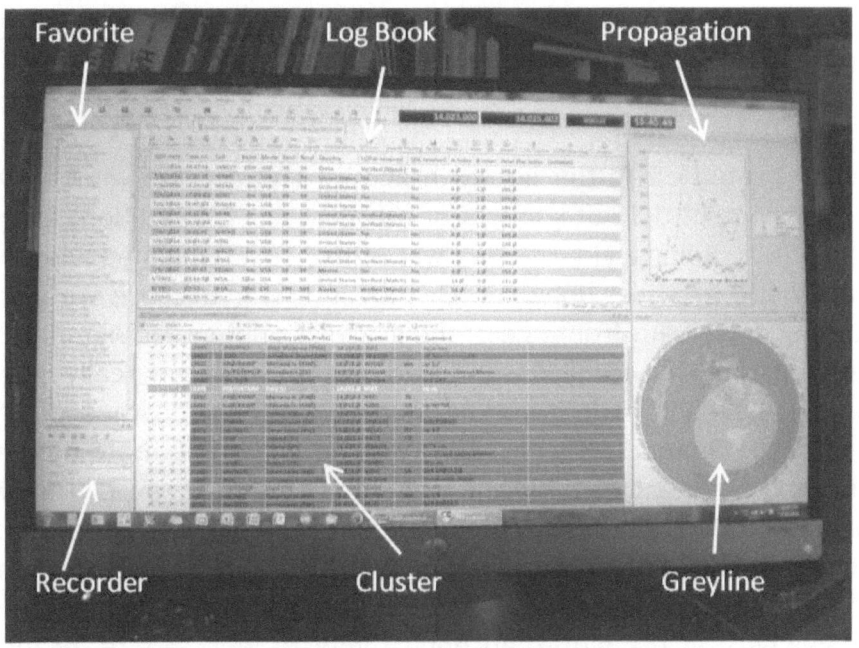

HRD Logging program. B/W does not do it justice. (G. Drasch photo)

The ability to identify DX, or any station for that matter, does not end with the cluster. How about an automated type of cluster? Absolutely, a CW skimmer, more commonly referred to as just a skimmer. A skimmer that can decode RTTY is known as an RTTY skimmer. A multichannel decoding program that decodes signals across a receiver's bandpass is the brains of these skimmers. It is specifically looking for CQing, along with the station's call sign calling the CQ. It uses a waterfall display, showing multiple traces which are then labeled with the station call signs. I view a skimmer as a robotic spotter. In contrast to a cluster, no person is involved in hearing or working a station, and then spotting it. The station is electronically detected and displayed robotically. After I used the cluster for a while, I discovered that not all the information is accurate. The case is, humans are entering the data, and that is why I have a tendency to rely on the accuracy of a skimmer a little bit more. As the cluster, any ham may own a skimmer, but instead of simply placing a PC on the Internet running a node program, he will require a dedicated receiver, antenna, and computer with the appropriate software.

When I returned to the hobby in 2010, I had no clue what a DX-pedition was. I did not know of or recollect any DX-peditions in the mid-1960s. Being a kid, maybe I was already overwhelmed with the hobby and was just unaware. Or maybe it was because we were turning downward from one of the greatest solar cycles when I first obtained my Novice license. I had just missed the really good propagation and hadn't worked much DX. When I learned of DX-peditions, I immediately wished I were younger, had more money, and had no heart condition. I made an attempt to explain a DX-pedition to Jeff, a non-ham friend of mine, at lunch one afternoon. I vividly remember how humored he was by it all. I began explaining that the sole purpose of a DX-pedition is to provide radio amateurs, everywhere in the world, an opportunity to work a rare location where other radio amateurs do not normally reside. Humankind recognizes 196 countries, but not all of them have resident ham operators. Having said that, the ARRL recognizes 340 entities, i.e., countries and islands combined, with which an operator can potentially work. So, how can anyone have a QSO with an entity if there is no ham there? Ah, that is where the DX-pedition enters the scene. A DX-pedition is comprised of radio amateurs, usually 10-20, who may or may not already be friends, or maybe they are acquaintances through various DX clubs. They each put in $15-20 thousand dollars of their own money, and then plan and propose a trip known as a DX-pedition. The destination will be some rare place ham radio operators have not been able to contact. Why is that?

The most likely reason is that the only inhabitants are penguins, rare birds, flora, bugs, or animals near extinction. It might be an island with just a few people and an abandoned WWII airstrip on it. To make this happen, the DX-pedition charters their own transportation via air and water, even chartering helicopters and specialized ships when necessary. They bring their own shelter, radio equipment, towers, antennas, generators, fuel, food rations, and drinking water when necessary. Sometimes, the governing body of the island will require them to transport all their garbage and waste back out. The cost of these DX-peditions can be astronomical; much more than the 10-20 hams can come up with on their own. Depending on the destination, the costs may go anywhere between half a million, to
three-quarters of million dollars. This is where the DX clubs and operators pitch in dollars to help fund this craziness. The Northern California DX Foundation, Metro DX Club of Illinois, and Greater Milwaukee DX Association, along with many many others from around the globe are examples of such. Another source of revenue for a DX-pedition comes from QSL card requests by those who made on-the-air contact with the DX-pedition. The money collected from QSLs is relatively small but never the less helps offset some of the costs incurred. Radio, antenna, and cable manufacturers might make monetary donations in addition to providing equipment.

Okay, back to Jeff. This is the kicker and the reason he was laughing so hard. The duration of a DX-pedition is relatively short, lasting a week or two, then packing up and returning home. Jeff says to me, "but they leave the operation, the radios, and antennas for others to use after they're gone, right?" I replied no! Jeff, said, "You mean it's just for making contacts; nothing else, they pack up and go home?" My reply was, "yup!" He was laughing so hard, I thought he was going to choke on his Belly Buster Burrito! I asked, "So, is it really any worse than a group of rock climbers paying outlandish amounts of money to climb Mount Everest?" Although there have been a few deaths on DX-peditions, I deem the odds of survival to

Gary L. Drasch

Certificate sent to the supporters of the unsuccessful Bouvet DX-pedition. (YB2TJV, D. Hidayat photo)

be much better. Plus all the radio amateurs on this sphere, rather than a select few, benefit from it all. It's a win-win! During the late 1920s and 1930s,[11] there were expeditions, but they were primarily geographical and exploratory in nature, for example, Admiral Byrd's Antarctic expeditions. One or two radio amateurs traveled with an expedition to make long-distance communications available. In their spare time, these same radio amateurs communicated with other hams who wanted to work a new country. The voyage of the schooner Kaimiloa is another example. The owners traveled the South Pacific in 1924, and while they enjoyed the islands, the radio operator, a ham, had QSOs with other hams stateside and eventually mailed QSL cards confirming the contact. Radio amateurs participated in geographical expeditions which were resumed after World War II. The Gatti-Hallicrafters Expedition to Africa in 1948, with Bob Leo, W6PBV, is an example. Although it was an expedition, it was not a DX-pedition for the specific purpose of activating an entity for radio amateurs. It was more of a Hallicrafters promotion touting their equipment after the war. As far as I can tell, the first DX-pedition as we know it today was done in 1948 by Robert W. Denniston, W0DX (SK), utilizing the call sign of VP7NG. It was to the Bahamas and was called "Gon-Waki." Denniston, who was an ARRL past president, is the one who created the concept of traveling to remote locations for the sole purpose of activating an entity. His call sign of W0DX is now that of the Caribbean Contesting Consortium, a memorial station in his honor. Later, there were adventures' such as Danny Weil, VP2VB (SK). Danny began traveling in 1955, often solo, to multiple locations in his sailboat named the YASME[12] (Yasme means, "To make tranquil" in Japanese). He hauled his radio gear along and provided many DX-ers with a new entity. Later, in 1959, The Yasme Foundation was created and led by Dick Spenceley, KV4AA (SK). The purpose of the foundation was to raise money to support additional DX-peditions by Danny until his DX-pedition retirement in 1963. Lloyd Colvin, W6KG (SK), and Iris Colvin, W6QL (SK), decided to pick up the Yasme DX-peditions and continued well into the 1990s. Currently, we have websites dedicated to DX-peditions such as https://dx-world.net/, and https://dxnews.com/. And what would DX-ing be without a DX-pedition calendar? You bet! You can find it at http://www.ng3k.com/Misc/adxo.html.

The ARRL introduced the DXCC (DX Century Club) program in 1937.[13] DXCC is an award available to those who have worked and confirmed 100 countries using different modes and/or bands (because islands count, it should be 100 entities rather than countries). Additionally, the award can be earned on each band and mode, or a combination thereof. The program was discontinued during WWII while the ham bands were dark, but started up again after the war ended and continues today. DX-ing started easy for me and then became harder as I accumulated increasingly more entities. I managed to obtain my DXCC using mixed modes, i.e., a combination of phone, CW and digital in about a year. It took the next five years to achieve it using every mode.

So, what is the big deal in chasing DX, especially after one receives the DXCC award(s)? Is that not the pinnacle? We know our station is working, why keep chasing? Each of us has a different reason, but for me, it all started with a banquet dinner I now attend annually. It was at the W9DXCC convention, an annual DX convention sponsored by the Northern Illinois DX Association in the Chicago area. After dinner, the host invited the attendees who were hams to stand up. He then said, "All of you who have worked more than 100 entities, remain standing." The people who hadn't needed to sit down. It was *"The Last Man Standing"* of DX! He kept going by increments of 25 until he hit 300, and then to 301, 302, and 303 until he reached 340 (340 is the maximum). I only had 126 entities under my belt at this time. It was humbling to sit down so soon, but it was likewise a revelation to see how few remained standing as the process proceeded. It was down to two men who radiated the feeling of standing forever; Bob Kelley, W0BW (SK), and Joe Schroeder, Jr., W9JUV (SK), with W9JUV[14] winning. It was awesome! Both of these gentlemen were much older than I. Later, another ham said to me, "You know what the secret to winning is? You simply need to live long enough!" Other DX associations around the country hold similar events every year. Try to attend at least one sometime. You will truly enjoy it.

The ARRL *"Honor Roll"*[15] is a prestigious award, which requires an applicant to be within the top 10 of the total number of DXCC entities available. As an example: An operator needs to confirm 330 entities out of the 340 available. (The number of entities available is determined by the ARRL, and periodically changes.) To me, the Honor Roll is the true pinnacle of DX-ing, and hope to realize it during my lifetime.

But wait, there's more! One day Lyle, WE9R, mentioned a DX-pedition that was active and said, "I need them on 80 and 160-meters." My reply was, "but you already worked them." He explained that he was filling band-slots. "'Band-slots?" I said. He replied, "Yes, that is when you try to work every entity on each band." I'm thinking, now that's something to do! (Remember—I am retired.) The ARRL refers to this slot thing as the "DXCC Challenge Award."[16] The person confirming at least 1000 band-slots, then 1500, 2000, 2500, and ultimately 3000, takes the award home. The score is determined by adding the number of entities he or she has on each of the HF bands plus 6-meters. For example, if I had 223 confirmed on 20-meters, 183 on 17m, 234 on 15m, and 186 on 10m, I would have a score of 826 band slots, or, as the ARRL refers to them, "band points." The person, who hits the 3000 mark, when divided by 10 bands, has worked and confirmed 300 entities per band!

I view DX-ing as a game or sport. Not necessarily a game of chance, but more of a mental and endurance game. A certain amount of technical competency, electronic, and electrical construction skills are involved. I draw an analogy to that of auto racing; an engine is built, a car is assembled, a driver needs to learn how to drive it, and then the team hopes he doesn't get hit or run off the track. My DX-ing sport has similar variables, e.g., the station I created and assembled, the antenna system I designed and erected my operating technique and skill, and the luck of good propagation. The operator, utilizing what he can afford, must develop the skill and technique to make that DX contact he's chasing. I was surprised when I visited the shack of a fellow member of the Greater Milwaukee DX Association, K9ORN. Brian is presently at 313 entities worked, and accomplished that using primarily vintage gear which he purchased as a young man. His antenna system is nothing fancy, it is a single Stepper IR vertical. This is a case where Brian knows how to use his gear AND how to work the pileups. Some of the best DX-ers I have met use rather old equipment, and the reason they are successful is that they know how to get all they can out of it. They furthermore know how to be found in the pileup.

I referred to the pileup earlier. What is it, and how does one work through it? As I explained earlier, nearly all DX stations work split, meaning they transmit on one frequency and then listen on a different one usually above, or sometimes below. The key to knowing if the DX is running split or not is to listen to them. If I hear the DX give their call and say, "up," I know they are working split. Pretty simple! The reason he is split is that if everyone was calling him on his frequency, there would be so many—calling continuously—that no one would hear the DX reply. All those stations calling, hundreds at times, are referred to as the *pileup*. If the DX station is popular but not necessarily rare, he might listen on only one frequency above his, and say, "Up 5," meaning he is listening 5-kHz up. If he is super rare, he will be tuning back and forth between 5 and 10-kHz up from his transmit frequency. A good DX operator says, "Up 5 to 10." I have a hard time to keep from laughing when listening to a pileup. I just find it hilarious at times.

I invited a friend to attend a DX club meeting with me one night. He replied, "You know, I'm not that interested in DX-ing." I call and call and they never hear me." What he was really saying is he does not know how to be heard in the pileup—even though he owns a radio with dual receivers. You might ask, "what does a radio with dual receivers have to do with breaking through the pileup?" Everything! The lion's share of transceivers has dual VFOs (variable frequency oscillators), A and B, meaning we may listen to one frequency on VFO A and transmit on a different frequency with VFO B, or the reverse. What we cannot do, is listen to both A and B simultaneously. To work DX which is split, we are going to want a radio in which we can listen to both the A and B VFOs at the same time, meaning two receivers within the same transceiver. This type of transceiver which has two receivers, not only has the ability to transmit on VFO B, but it can additionally listen to that frequency. Do you see where this is going? If we were to park our transmitter, VFO B, on a frequency in the pileup and wait for the DX station to find us, we will be waiting for a long time. Or maybe, as with my friend, we may never be called at all. However, if we can hear which station the DX station worked last, we would be close to where he is listening. Once we have that radio with dual receivers, we continue listening to the DX station on his frequency, VFO A, and tune our VFO B into the pileup, and listen for the last station the DX worked. The DX station might remain listening there for a minute or two and work a couple of stations on that frequency. Alternatively, he might move up or down a kHz or two at a time and pick up another caller. If that is the case, we tune the second receiver, VFO B, (keep in mind, our transmitter is tracking our second receiver) either up or down and we try to get barely ahead of where he last worked a station. Immediately, we have an exceptional chance of being his next contact. I find this to be the real fun in DX-ing, the mental game of what the DX station is doing. Is he moving up or down, does he come all the way back down to the bottom after he gets to the top of the pileup and starts over again moving up, or does he start moving back down after he gets to the top? Does he jump from the top to the bottom and never listen in the middle? Maybe he randomly moves about in the pileup, up one, down two, down another one, up four, and back down three and etc. That is the worst scenario—it is impossible to determine a pattern because he does not have one.

You have probably been thinking; how does a DX-er decipher, or distinguish, the signals between the DX station and the station he contacted a moment ago? The standard way is to use cans, otherwise known as headphones. If the radio has dual receivers, it gives us the ability to split the VFOs on the headphones. There might be a couple of different configurations available depending on the brand of radio. The common way to listen to the DX station on VFO A is with the left ear, and the pileup, VFO B, on the right ear. That might sound crazy but it works well once you have done it for a while. My hearing in my left ear is impaired compared to my right, and all of a sudden, I realized I was starting to have trouble hearing the DX on VFO A. As with most dual-receive radios, I was able to adjust the volume of the two VFOs separately on my radio. Although I had been doing that for some time, I continued to have trouble hearing the DX. Then one day while going through the manual I uncovered a setup where the operator can listen to VFO A

Gary L. Drasch

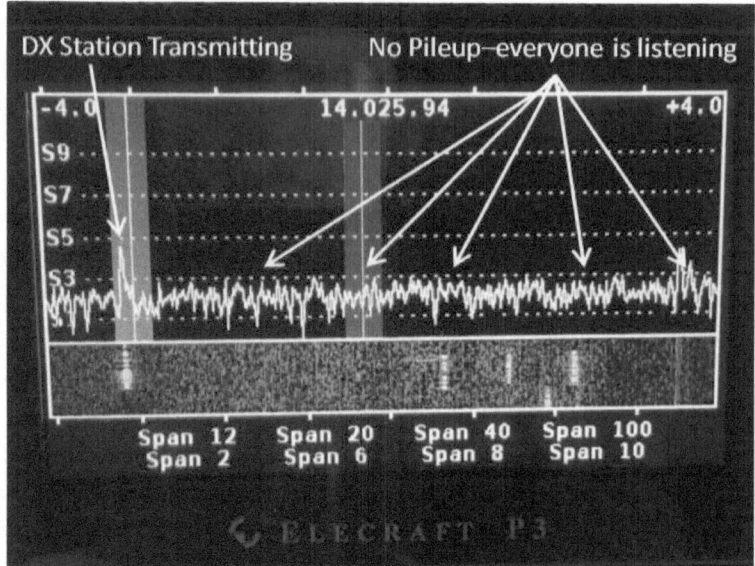

Panadapter display: Here the DX station is transmitting and the complete pileup is listening. (G Drasch photo)

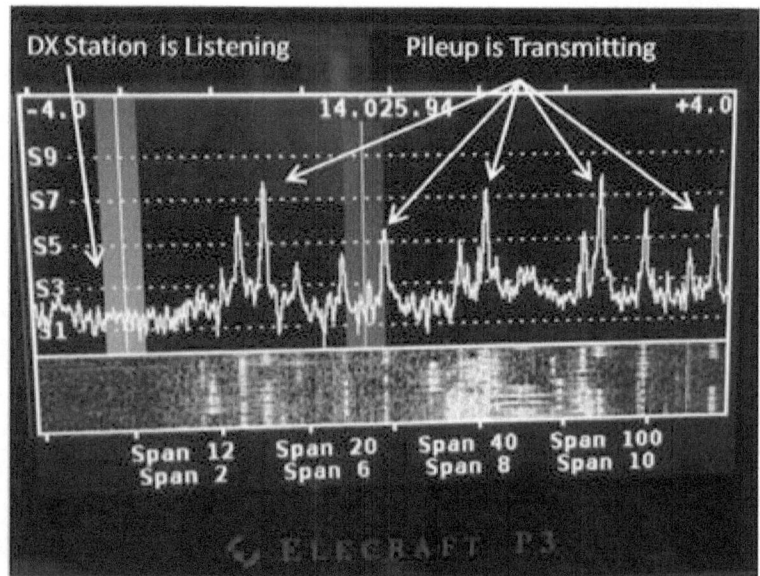

Panadapter display: This time the DX station is listening and everyone in the pileup is transmitting. (G Drasch photo)

with both ears, and configure it so VFO B is only heard on the right ear. Perfect! By the same token, it can be set up in the reverse with VFO A solely on the left and VFO B on both. If we are not wearing headphones, most dual-receiver radios allow us to use a pair of speakers in the same manner.

Now that we know how to find where the DX is listening, it is time to give him a shout. But before we do, can we hear him? The DX that is. We know where he is listening, but is the DX readable? We do not want to call him if we cannot hear him well enough to know if he is replying to us or a different operator. In nearly all cases, the DX station can hear better than domestic stations because they do not have the noise as we do stateside. It is possible for them to hear us and we are not able to hear them. If we call and are unable to hear them, we are wasting their time. Okay, assuming we have a solid copy on them, we will move forward. While tuning in the pileup, we likely found operators giving their call sign two or three times, right in a row, almost connected together. They are still calling while the DX station is already replying to a different caller! How is the DX station going to hear them if he is transmitting and calling a different station? What the DX station is listening for is a single call, not multiple calls, from a station. What we need to do is find where we want to transmit and give him our FULL call once, and then listen. (He is NOT interested solely in the suffix of a call sign—it slows things down—he wants the full call.) If he remains quiet, we call him again, once! If he comes back with a different call sign than ours, we do not transmit unless we want to antagonize him. I hear this happening all the time. If he is asking for W6XYZ and we are N3DAH, why would we call him? He is not replying to us, so we wait until he does. The DX asks for one specific call sign and three or four others all reply—not cool. Wait for the DX to finish the contact. When we hear a "thank you" come from the station worked, we listen to ensure the DX is not transmitting, and then we call him again, only once. We need to be in the habit of calling only ONCE, then listen, call once, and then listen, and so on. This is likely happening as we are following other callers with VFO B. I have frequently gone into huge pileups and made the QSO in one or two calls. It is not going to happen all the time but our chances of making a contact are much better than not. At the point, the DX station calls us, we reply by saying, "59 thank you." In CW, it is 5NN TU, and the TU is sent so fast that it sounds like an "X." (The "N" in 5NN, rather than 599, is referred to as *cut numbers*. I will talk about them later.) Consider this for a moment, the DX is trying to accommodate literally thousands of other hams, the likes of you and me, in a finite amount of time. Consequently, they want each contact to consume the least amount of time possible. Of late, it is becoming common to reply only with 5NN and skip the TU simply to shorten things up and use less time. Time is that important!

Prior to 1961, DX-peditions would send and receive on the same frequency, and if they were operating CW, they would send 599 and therefore receive a report of 599 too. The concept of operating split and using 5NN was the brainchild of Don Miller, W9WNV.[18] Don had once said, "If I ever have an opportunity to be on a DX-pedition, I'm going to do things differently." His chance arose while serving as a physician in the military during the Korean War. He was licensed as HL9KH while in South Korea and came up with the idea of shortening the time spent on each QSO, hence the current standard of operating DX.

Operators who work the DX-pedition more than once on the same band, in the same mode, are doing nothing more than wasting the DX-peditions time. They need to stop that. We do not do it unless we know for certain that we are not in the DX-peditions log with our first contact. I have met a few individuals who work them twice to ensure they got them; a terrible misuse of the DX-peditions time and the potential of robbing a fellow amateur of working an all-time new one (ATNO). Here is a tip. To ensure we worked him on a particular band, we work him again but we use a different mode on the same band. If I worked him using CW, I then pursue him using SSB or maybe RTTY. That way, I can reasonably count on having them on that band. Many DX-peditions upload their logbooks via satellite to Club Log or their own website where we can check and see if we are in fact in their log. If we do not find ourselves in it, we need to try them again. Who knows, we may have worked *Slim*! (More on Slim later.)

What about a panadapter or a bandscope? Why use one of those? The terms panadapter and bandscope are used interchangeably to describe a radio spectrum scope. Marcel Wallace[19] invented a panoramic spectrum display in the 1930s which provided radio operators a graphical display of signal amplitudes versus frequency. Today, the device visually shows signals (stations) within the RF spectrum. For example, we may look at a bandwidth as little as 2-kHz wide and up to 200-kHz wide with a panadapter. This gives an operator the ability to see each signal and its amplitude (strength) within the bandwidth chosen. We are able to see a complete pileup. As a result, when the DX station replies to a specific call sign, we will see the calling station popup on the display when he replies. We can see where the DX has been listening and in what direction he is moving. What hinders this method is when people who are NOT being called also reply. Instead of one signal popping up as it should, we might see three or four. Watch for signals that popup in the exact same place each time the DX calls for a reply. They are NOT the ones being called. Those being called are in a slightly different place each time. Those ARE the ones we want to follow with our VFO B. Bandscopes have become so much of the norm that they may have lost some of their advantages. That is partly due to more radios having them built-in as the one pictured below. And let us not forget SDRs (Software Defined Radios) whose receiver display platform is based on a panadapter, and displays multiple slices of the RF spectrum on a PC monitor at one time. Some hams have even created their own bandscopes by connecting the output of the 1st I.F. of their radio through an SDR dongle, software, and a PC.

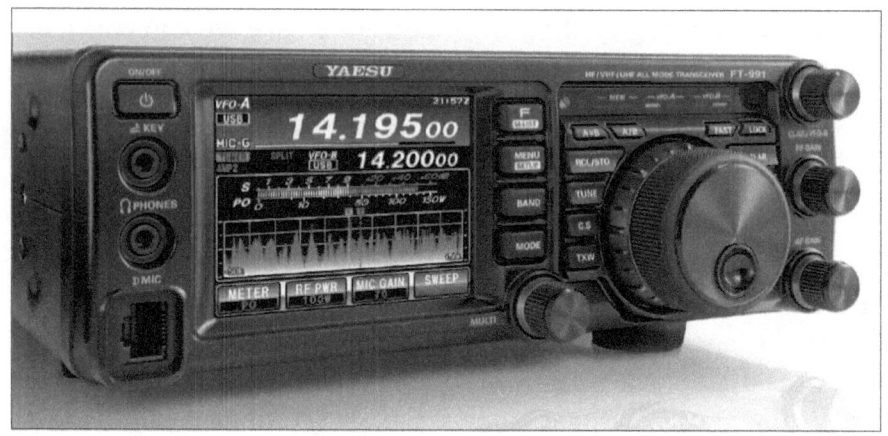

An example of a built in Bandscope— but single receiver. (Yaesu photo)

A few operators might be afraid to try CW for DX-ing because the DX stations are transmitting at maybe 24 *words-per-minute* (wpm) or faster. Do not let that scare you. You do not need to copy any text at that speed other than your own call sign. After all, you are not going to have a full-blown QSO—it is a 5NN TU, and you are done. I am good at about 17-wpm in a standard QSO, but when working a DX station, I can pick my call sign out at about 30-wpm. Go practice. Look for free CW practice programs on the Internet. The other thing you need to practice is sending your call sign and 5NN TU, at about 22 to 24-wpm. I try to adjust my transmitting speed to that of the DXs. The only caveat is that even if he is at 30-wpm, I remain at 24-wpm.

A little more frequency spectrum is adding to the DX experience nowadays, i.e., we have some additional bands to use. The WARC bands, 12, 17, and 30-meters were made available to us after a 1979 conference in Geneva, Switzerland known as the World Administrative Radio Conference[21] (WARC). The bands were available to use in the early 1980s. Referring to the ARRL band plan,[22] you will notice they all have a narrow bandwidth which is 100-kHz or less. Because of this, it was agreed not to permit contesting on them. This created a nice place for non-contesting amateurs to meet and operate without the interference of contesters. It was further decided that phone communications would not be permitted, and no more than 200 watts P.E.P. (peak envelope power) allowed, on 30-meters in the United States. This created a terrific place for the CW and digital/data boys to operate.

The 60-meter band was introduced in 2002,[23] but only a few countries were allowed to use it. It is not considered a WARC band because it had not been part of the conference. Only Denmark, Finland, Iceland, Ireland, the United States, and the United Kingdom had the privilege of operating on 60-meters. Later, in 2015, the ITU World Radio Communication Conference approved a worldwide frequency allocation of 5351.5 – 5366.5 kHz to be used in the ham bands. Referring to the ARRL band plan, you will see the band is broken into five channels, and the power is limited to 100 watts P.E.P. Moreover, because the band is shared, the United States requires us to use *USB* (upper-sideband) for phone communications to accommodate intercommunications with non-amateur services, if necessary.

What is a vanity call sign and how do we get one? In 1978,[24] the FCC decided to grant, or let's say create, vanity call signs. They are referred to as 1x2, 2x1, or 2x2 calls, e.g., K2HB, NM4P, and KA1AH respectively. Both DX-ers and Contesters alike, often have a vanity call in order to shorten things up; to make it easier to work DX contacts, or faster to make contest contacts. To obtain a vanity call sign, we need to have a call sign to trade-in. A Technician class, when originally licensed, is issued a 2x3 call sign by the FCC, such as KC9XXX. He or she is unable to choose that call sign. Once they have it though, they may trade it in for a 1x3 vanity call sign. The same holds true for a General class licensee. Advanced class licensees may apply for a 2x2 call sign and Extra class holders may apply for 1x2, 2x1, or 2x2. RadioQTH.net located at http://www.radioqth.net/ is a call sign search tool that lists various call signs and availability. We can apply for the call sign we are interested in by using the FCC Universal Licensing System (ULS). On September 3, 2015, the FCC did away with the regulatory fee only to reinstate it on April 19, 2022. The sole reason I have not applied for a vanity call is that "DJT" has been with me since I was 12 years old. First as KN9DJT and then K9DJT. It's just a nostalgic thing...

The CW Contesting and DX station of W0GXA (W0GXA, R.C. Lee photo)

Chapter 4

Propagation

When I was 13 years old, in my mind, the bands were either open or not. I did not study propagation nor thought about it a whole lot. I was utterly unaware of the *solar cycle* at the time. It was probably a good thing too because cycle 19 was on the way to bottoming out. Looking back into my tattered light blue 1962 ARRL Handbook,[25] I found only two paragraphs referring to cyclic variations, and I don't remember reading them. By the condition of the book though, I must have come across them at some point. Or maybe because I didn't have any control over the band conditions, I felt it wasn't worth the bother. Propagation is truly a study within itself. We have so many more tools available to help us today than when I got started. You may want to do some general searching, or you can go to the ARRL website at http://www.arrl.org/propagation-of-RF-signals. It contains an entire list of valuable articles. Besides articles and tutorials, there are propagation charts and predictions. And no doubt, many books are available on the subject.

Years ago, the only way we knew if a band was open, meaning there was propagation, was to listen and tune around. Plenty of times, I did not bother to call CQ because I didn't hear anyone. Thinking back, that wasn't too smart. If everyone else who was listening thought the same as I, no one would know if the band was open or not. Then again, being only 13 years old, I wasn't thinking that far.

To get a better handle on propagation, rather than listening for other hams, The Northern California DX Foundation (NCDXF), came up with an idea and created the first beacon network at and around the frequency of 14.100-MHz. The year was 1979.[26] It was decided the network would be international in nature, having locations around the globe, with each site transmitting a signal at regular intervals. Each site location can be identified by its call sign in Morse code. In this way, the beacon network allowed radio amateurs, no matter where they lived, to assess the current condition of the ionosphere and the propagation of radio signals to any location on earth. The first beacon was designed and built by Jim Ouimet, K6OJO (SK). He did this because of a suggestion by O. G. "Mike" Villard, Jr., W6QYT (SK), who was a professor of Electrical Engineering at Stanford University during the time. Mike was the same person who introduced SSB to the amateur bands in the late 1940s. The FCC licensed the first beacon as WB6ZNL/B (the /B indicating a beacon). The network consisted of nine frequency-sharing CW beacons and had been in operation for almost 15 years when the *International Amateur Radio Union* (IARU) proposed a larger network. They were so impressed with the system that they proposed a network of beacons that would operate 24 hours a day. In addition, they wanted the network to operate on five bands with the transmissions shortened from a one-minute message every 10 minutes, to one every 10 seconds. This enabled listeners to monitor all the beacons in three minutes. The transmission of the beacon call sign is sent at 22-wpm and then followed by four one-second dashes. The call sign and the first dash are sent at 100 watts. The power then decreases to 10 watts, one watt, and then 100 milliwatts for the remaining dashes. After each 10-second transmission, the beacon steps to the next higher frequency band. The locations are as follows: New York City, Canada, California, Hawaii, New Zealand, Australia, Japan, Russia, Hong Kong, Sri Lanka, South Africa, Kenya, Israel, Finland, Madeira, Argentina, Peru, and Venezuela. The frequencies of each location, along with more information, are available at http://www.ncdxf.org/beacon/index.html. Because of sufficient worldwide funding, the NCDXF/IARU Beacon Project was later expanded to 18 beacons in 1995 and has been operating ever since. My logging software allowed me to add the beacon frequencies to my favorite frequencies list. When I want to see what type of propagation is going on, I simply double-click a beacon in the list and the program takes the radio there.

Another new propagation network emerged not too long ago. It is called the Reverse Beacon Network[27] (RBN). This network, instead of transmitting a signal regularly, listens to the bands using a CW skimmer and reports what stations it hears. The RBN includes hundreds of monitoring stations worldwide. Unlike the NCDXF/IARU project, the RBN is strictly volunteers who are willing to set up and maintain a monitoring station at their own expense. The resulting information is displayed on a website having the appearance of a telnet cluster. It shows what station is heard by what skimmer, what frequency they are transmitting on, a signal-to-noise ratio level in decibels, and at how many wpm they were sending. Try it at http://www.reversebeacon.net/main.php and call CQ using CW. You'll be able to see where you are being heard on the map.

Okay, when should you operate? My normal response is, "anytime you want!" It's like fishing—anytime is a good time! But as fishermen know, there is a difference between "fishing" and "catching." The same holds true when trying to work DX. Anytime is a good time to DX, but if you are interested in catching more DX, you might want to focus during the grey line—that is the time just before sunrise, and just before it goes down. The grey line is a ring around the Earth separating daylight and darkness. The D-layer of the atmosphere, which absorbs HF signals, has not built up yet on the sunrise side before it disappears on the sunset side of the grey line. As a result, DX communication is exceedingly good throughout this window. Many map programs show the grey line live. My logging program provides such a map, which I keep displayed while operating. There is a good *QST* article, July 2010,[28] by Steve Sant Andrea, AG1YK, titled, "When Should I Operate?" explaining this phenomenon. (Old *QST* articles are available for download to members of the ARRL as a .pdf at http://www.arrl.org/arrl-periodicals-archive-search).

I confessed earlier that when I first got started I was unaware of the solar cycle, i.e., the 11-year change in the sun's activity. When the levels of solar radiation and expulsion of solar materials interact with the earth's magnetic field, radio propagation is drastically affected. The number of sunspots and flares all have a positive effect. The more sunspots during a year, the better it is. The recording of sunspot activity first began in 1755.[29] That means the current solar cycle, number 25, is the 25th cycle since then. I was fortunate to have re-entered this hobby during the upswing of the previous cycle. It began on January 4, 2008, but first started generating reasonable activity in 2010. I thought the band conditions were fantastic but later learned cycle-24 went down in history as a cycle with the lowest activity ever. Nevertheless, I spent the majority of my time on the higher bands because they were so good. A savvy ham will always work the higher frequencies while the cycle is high. So, what did I do at the bottom of the cycle? I focused on the bands which worked—the low bands, 30, 40, 60, 80, and 160-meters. It does not mean I gave up on the higher bands, it meant I monitored the beacons and hoped for some openings. I further made use of the *weak signal* (WSJT) digital modes, i.e., FT8 and FT4. I also started exploring the use of meteor scatter on 6-meters for the first time.

In the end, it comes down to another variable radio amateurs need to deal with when chasing DX. If it was easy, all hams would be on the "DXCC Honor Roll" and the fun would be out of it. As far as propagation; we can study it, track it, plot it, and predict it, but we cannot control it. Because of that, we get on the air and operate as we always do. Hey, maybe I wasn't all that wrong at 13 years old!

Chapter 5

Contesting

Radio contesting, sometimes referred to as *radiosport*, received its start back in the 1920s.[30] This happened to be when radio amateurs in the United States and Europe were attempting to communicate across the Atlantic Ocean. At the time, more radio amateurs were trying and becoming successful in communicating over large distances. This was the advent of DX-ing! After communications were established across the pond in 1923, the testing and attempting of transatlantic communications continued and soon became an annual event. The ARRL was the organization that was promoting and publicized these tests. Later, in 1927, the ARRL created a new format for the annual event by asking stations to try to make as many contacts as they could with operators in other countries. This milestone soon became known as *The 1928 International Relay Party*[31] and thus the very first ham radio contest. It was such a success that the ARRL began sponsoring it annually through 1935. In 1936, a name change was in store, and what emerged was the *ARRL International DX Contest*. It continues today, retaining the same name. Contesting continues to be part of the hobby where radio amateurs around the globe try to make as many contacts as they can with one another while following certain parameters. For example, besides exchanging each other's call sign, they might be required to exchange a signal report, a name, a state, or maybe a serial number, such as, "You are contact number 412." Contestants also have the possibility of increasing their scores by obtaining multipliers. A multiplier might be working different counties, states, entities, or maybe call sign prefixes. As with any type of competition, there is a specific amount of time allotted.

Gary L. Drasch

Contests are sponsored by a variety of organizations ranging from the ARRL, IARU, *CQ Magazine,* and radio clubs in the United States and the world. The awards presented by the sponsors in the various categories range from certificates to plaques.

There's a love-hate relationship among radio amateurs when it comes to contesting. Some hams love it but some absolutely hate it. The rationale is, contesters occasionally interfere with a regular scheduled QSO or informal net. They don't do it intentionally. It's just something that happens because of the pure volume of contesters on the air at the same time. Many contests now have suggested frequencies to help relieve such conflict. The other thing available to the *haters* is the WARC bands. It was decided not to allow contesting on the WARC bands because of the limited bandwidth of each, but additionally as a frequency spectrum for those who are not interested in contesting. This provides the non-contester a place to operate without the possibility of contesting QRM.

Take a moment to peruse the WA7BNM contest calendar at http://www.hornucopia.com/contestcal/contestcal.html, and you'll likely find a contest occurring every weekend. There is a contest for you no matter what duration of time or mode interests you. Long ones go for 48 hours, to short ones lasting 12 hours, and even mini ones lasting only 30 minutes. There are CW, SSB, digital and mixed modes. For reasons unknown, I did not get into contesting until recently. Maybe it was due to a conversation I had with a friend of mine. He said, "Unlike DX-ing, where you work a finite set of entities, with contesting, you start over each time. You begin at zero and build from there." As there are DX clubs, there are *contesting* clubs. There are also contesting stations. Large ones, specifically designed for "Multi-Multi" contesting, have three to four operating positions all going at the same time. Or how about stations having the capacity for a separate operator on each band? Yup—all running simultaneously. And undoubtedly, there's a magazine, the *National Contesting Journal* (an ARRL publication).

I joined the Society of Midwest Contesters club in 2016 because several members of my DX club belonged as well. DX-ing and contesting go hand-in-hand. It is about building a station, developing a high-performance antenna system, and honing the skills to compete among friends and other clubs. Plus it is a means of working countless DX and/or states on a weekend. In the last 160-meter CW contest, I managed to work all states except for four. Furthermore, contesting sharpens our operating skills. My CW copy speed has gone up, and it is easier for me to copy call signs when using phone. Once you have contested for a while, you will find that a rhythm develops while making all those valuable Qs. It's fun! There was a gentleman (I am unable to remember his name) who was the keynote speaker at one the of W9DXCC banquet dinners I attended. What he said surprised me until I thought about it. He said, "It isn't a question of our nation's communication systems

K9CT Contest station during the CQ World Wide Competition. (K9CT, C. Thompson photo)

failing someday. It is a question of when." He went on to say, "There are all types of radio amateur emergency nets in existence, but I believe the *contesters* will be handling the majority of emergency traffic, simply because of the out-and-out volume of it." The more I thought about it, the more I started to think he was right. It is fairly evident during Field Day where there is an honest cross-section of operators. It's easy to differentiate the operators who are contesters compared to those who are not. Contesters have developed a technique to communicate with hundreds, if not thousands, of other contesters accurately and quickly.

A contester does all he can to maximize his number of contacts. Seconds count! Abbreviations such as "CQ test," instead of "CQ contest," are used frequently when operating. If the name of the contest has an abbreviation that is shorter than "test," she uses that. Another example of a true contester is he finds a frequency and parks there while he calls CQ (this is referred to as *running*). He will S&P (search and pounce —tuning and looking for others calling CQ) only after he feels nothing is left to be had on his frequency.

K9CT contest station during the CQ World Wide competition. (K9CT, C. Thompson photo)

He wears headphones and does his own logging on the PC. His communication with the other operator is short and to the point. When the exchange is made of the signal report and classification, he does not bother saying, "please copy" each time he provides his. *Please copy* is a total waste of airtime. He may or may not even say thank you at the end of the contact because of the amount of time it takes. The goal is to collect each other's information and move on to the next as quickly as possible. That's how to rack up a lot of points! A CW operator does the same thing as a SSB operator, but he takes it a step further. He uses what is referred to as *cut numbers*. When sending the exchange of the signal report and classification, instead of sending 599, he sends 5NN, the same as working DX. If you are a non-code person, "9" is sent as DAH DAH DAH DAH DIT. Now compare it to the much shorter "N" which is DAH DIT. When it comes to the serial number, the true CW contester has two more cut numbers she uses. They are for numerals 1 and 0. He sends an "A" for 1, and for the numeral 0, he sends a "T." Again, in code, the difference being 1 is DIT DAH DAH DAH DAH, and an "A" is DIT DAH. The 0 in code is DAH DAH DAH DAH DAH, and a "T" is simply a DAH. The origin of sending a T for 0 began with the American Morse code,[32] otherwise known as the Railroad Morse during the mid-1840s, when code was still being sent over the wire. The practice was to send an extra-long dash for a 0, but because of electronic keyers, ham operators began sending a single dash. The real purpose of shortening things up is all about time. I know, it is barely a second or two difference per QSO, but all those seconds add up.

The code speed during a contest varies. If I am doing a search and pounce, I try to match the CQ-ers' speed when calling them. When using a contest logging program, I am not using my key or paddles to transmit. I select the speed I want to send at within the logging program, and when I press the appropriate function key, it sends the code for me. As far as copying him, I probably heard his exchange once or twice from his previous contact and already entered the data. S&P takes more time to make contacts because I am tuning around between Qs. Because of that, my copy speed is not as critical as if I were calling CQ. If I am running, or calling CQ, I normally do it at 20-wpm. I hear others calling CQ at 24-wpm and faster because they are usually able to copy faster than that. At 20-wpm, I have callers at 17 to 18-wpm and maybe up to 22-wpm, but in general, contesters try to adjust their speed to mine.

I admitted I am good for a conversation at about 17-wpm, but I can copy call signs at a much faster speed. The reason is that I practice copying *nothing but* call signs. Who would have thought of specific programs and smartphone apps to assist us in doing that? I have an app on my iPhone called the *Koch Trainer*. The Koch Trainer is based on the *Koch method*[33] of teaching Morse code. It was developed by a psychologist by the name of Ludwig Koch in the 1930s. It provides choices of listening to words, call signs, QSOs, and random letters all at various speeds. I likewise use two other programs on my PC. The first one simulates a contest that includes interference and static while it sends call signs. It is called *Morse Runner*[34] and was created by Alex Shovkoplyas, VE3NEA. I need to select a code speed that I normally use during a contest in the setup. The program then simulates callers at varied speeds near it. I type in the call signs as I copy them, and get graded on my accuracy at the end of the exercise. Another one is *RUFZ*,[35] which was created by Mathias Kolpe, DL4MM, and Alessandro Vitiello, IV3XYM. The RUFZ program is also used for call sign copy training. With this program, I first need to select a starting copy speed within its setup. Then, once I start an exercise, as I copy and type in the call sign, the program decides if my entry was correct or not. If correct, the next call sign comes a little faster. If I was correct again, the next call sign is still faster. The speed keeps increasing until I make a mistake. When that happens, it starts to slow down. Make another error and it slows down again until I get one correct. At that point, the speed starts increasing again. I regularly use the Koch trainer on my phone because it is always with me. I use the two trainers on the PC for a while before an up-and-coming contest. It's a good way to sharpen our skills!

If you want to get started in contesting, it is imperative to interface your radio with a PC. You then need to select a contest logging program. There are many: N1MM+, N3FJP, and WriteLog are just a few. I happen to like and use N1MM+. It can be found at https://n1mmwp.hamdocs.com/. This program is free, it works, has all the current contests embedded, and is updated on a regular basis. I dare say, it has pretty much become the standard among contesters. Once you have your radio and software interfaced, it is time to select a contest to enter. Contests require you to select a classification that fits your station configuration. You will need to understand various contest abbreviations[36] to do so. Here are a few:

MO-Multi-Operator
MS-Multi-Operator, Single Transmitter
MM-Multi-Operator, Multi-Transmitter
M2-Multi-Operator, Two Transmitters
SO-Single Operator
SOAB-Single Operator, All-Band
SOSB-Single Operator, Single-Band

The power category you want to run needs to be selected as well, e.g., HP-High Power, more than 100 watts, but not to exceed 1500 watts PEP; LP-Low Power, 100 Watts PEP or less; QRP-Very Low Power, 5 watts PEP or less, average output power. Another part of the classification setup is selecting either Assisted or Non-assisted. If you want to use the cluster or skimmer when operating in the contest, you need to select Assisted. Obviously, Non-assisted is for those NOT using any type of aide. The ARRL has a nice contest glossary at http://www.arrl.org/contest-glossary. You'll find a surplus of information there. You may have noticed classifications for the serious contesters which permit the use of two operators and two transmitters or more. These are used by the contest stations I mentioned earlier. What I didn't include in the list was the acronym **SO2R**. It stands for Single Operator Two Radios. There are hams in my contest club who are very good at this. For my part, I remain baffled by how they manage to do it. How this plays out is that a *single operator* operates two separate stations, on two different bands concurrently. He *runs*, calling CQ and answering callers on one band; on the other band, he *searches and pounces* looking for people calling CQ. He differentiates the two stations the same as a DX-er would; listening to one receiver in the left ear, and the other in the right ear. By entering the contest as a single operator, he needs to make certain he transmits with only one of the two transmitters at any given time. There is even special hardware marketed to the contester to ensure it.

The Internet managed to bring some niceties to the contesting game too. These days, contest logs are sent to the sponsor via email or uploaded through a website in a standardized format called *Cabrillo*. Also, there is no more waiting for my results to be published and wondering how I faired. I can go to a website at http://www.3830scores.com/ and compare my unconfirmed score with others. All I need to do is find the contest I participated in on the website, and fill in a form with my summary data of the contest, e.g., the number and type of contacts on each band, the number of multipliers, my power, and the classification. Without question, all this data is available from my contest logging program. All the scores are claimed and unconfirmed, but nonetheless, it is still cool to see my score among the others. If I'm in competition with a couple of friends, or between clubs, and have an Internet connection, I can share my log in real-time to http://cqcontest.net/view/readscore.php and compare scores live. Imagine how much fun that is!

Once you get into contesting, you will become more serious each time you participate. For example, I frequently practice my CW copying skills before a contest and set up my chosen contest within my logging program a few days earlier just to make sure the station plays as it should. No one wants to be fooling around trying to make things work at the time the contest starts. It is no different than preparing for a fishing trip; I want my rod and reel ready when the fish start biting.

I planned to do the *Wisconsin QSO Party* from my cabin in Forest County, WI, only to have it turn into a complete bust.

SO2R operating position of K3WA using the very latest technology. (K3WA, W. Axelrod photo)

And here is the station muscle! (K3WA, W. Axelrod photo)

Gary L. Drasch

The shack of K9SD. (K9SD, S. Effinger photo)

That happened even though I checked my stuff out before leaving on the three-hour road trip. I set up my notebook PC and radio combination at home, checking the headphones, voice-keyer, and footswitch, leaving nothing to chance. At any rate, that is what I thought. A long story short, my Carolina-Windom antenna at the cabin would not tune. Up till then, I had been happy to see it up in the trees when I arrived and never gave any thought to the transmission line being troublesome. Water made its way into a coax connector splice, froze, and broke it apart. I cannot tell you how heartbroken I was. And it was just a stupid seven-hour contest! A friend of mine had a mishap that created a similar result. He was ratcheting up his crank-up tower 30-minutes before the start of the World Wide DX Contest when the cable broke. The complete tower collapsed within itself. The main thing was he did not get hurt, but his beams and pride were. The booms of both beams, at the mast, were bent downward at about a 35-degree angle due to the sudden stop. More than a few days passed before he came out of his funk. I forgot to mention, he takes contesting very seriously!

Be prepared—contesting and DX-ing are equally addictive!

Chapter 6

QSL-ing

It should be of no surprise that an electronic means of confirming QSOs exists for all those awards we enjoy chasing. Although so much more practical and affordable, it is different from receiving that special QSL card in the mail. I look back to the 1960s and can see myself handwriting QSLs, placing the QSO information AND the operator's address on the same card. I then placed a three-cent stamp on it and dropped it in the mailbox. It resembled a postcard—having a great time, wish you were here—remember those? Well, if you enjoyed receiving those jewels in the mail you are still in luck. My friend Rudy, NF9V, likes collecting QSLs. Rudy has some special cards displayed in his shack, but in addition, he has QSL-sized plastic file cabinets literally filled with cards, all in alphabetical order. The exchanging of QSL cards is very much alive, except it costs a good deal more. In the past, if I sent a QSL to a contact in the United States, he or she happily reciprocated with one of theirs. Hams who now do are few and far between. The new norm is to send our QSL in an envelope, along with a self-addressed stamped envelope (SASE). At the time of this writing, the cost is 58 cents to send ours there, plus 58 cents for the return SASE, plus the cost of two envelopes. Ah, and not to forget the cost of the QSL card. I recently had 1000 custom cards printed for $80. Hence, to receive a wanted card, I am going to spend a total of $1.24. And that's domestic. Now consider obtaining a card from a station in a different country. I again send my card along with a self-addressed envelope (SAE), but instead of placing a stamp on it, I am going to enclose three "green stamps," i.e., dollar bills. Those three dollars are used by the recipient to mail his card to me. In some cases, the DX station might even request more money. Additionally, the postage of $1.30 is needed to send all the above to him. So, let me add this up. The cost of my card is eight cents, the dollars sent to him for postage is $3.00, and the postage to mail all this is $1.30. That comes to $4.38 to receive a single QSL card. There are times when I sure hope I am in his log and he doesn't use the money for beer. HiHi (laughing in Morse code). My original comment when hearing this from my friend Lyle, WE9R, was, "that is outrageous!" Lyle replied, "Hey, It's cheaper than smoking!" Neither one of us has ever smoked, but he was right. I have often rationalized expenses relating to my hobbies by comparing them to the cost of owning a boat. Not only is there investment in the boat itself, but there is insurance, license fee, fuel, maintenance, storage, slip, or launching fee. If fishing is of interest, there is the expense of all the tackle. Then the refrigerator needs to be stocked with beverages and food. And with all that, the usage is dependent on the weather! Hell, $4.38 isn't too bad for a memento I can reminisce by, collect, or use toward an award.

The QSL Bureau,[37] which is commonly referred to as just the *bureau*, carries on. It was originally established in the 1920s,[38,] and is a system of multiple clearinghouses specifically for QSL cards. They are located in various call areas of the United States, and similarly in other countries. In my area, the Northern Illinois DX Association (NIDXA) provides the service. One of the negative things about using the bureau is that the process is staggeringly slow. Sometimes I might need to wait a year or two before I receive a QSL card. Therefore, I only use it to reply to QSLs I receive via the bureau, or for cards, I want but am willing to wait for. Understandably, the bureaus need to offset their rising costs periodically. At present, I send $10 to NIDXA when needed. That is applied to my account and then drawn upon for the cost of postage to mail received cards at their location, the area nine bureau, to me. When my account gets close to being depleted, they will enclose a note along with my bundle of QSL cards to let me know. It's an inexpensive way to receive a bunch of cards. To send cards through the bureau used to be a little more expensive because of a flat $7.00 service fee. The ARRL has now dropped it and created new rates. The present-day cost through the membership QSL Services (bureau) is $2.00 for 10 or fewer cards in one envelope. For 11-20 cards in one envelope, it's going to cost $3.00, and for packages with 21 or more cards, $0.75 per ounce. But that does not include the postage from my QTH to the bureau. As you can see, the cost isn't bad. I can decide to do all my confirmations electronically, but there is a chance the rare DX stations do not QSL in that manner. Each QSO requires thought as to what type of confirmation is best. Do I want a card to add to my collection? Or maybe I want a card because the QSO had been meaningful. Maybe I only need a verification to put towards an award. It might not be entirely up to me, but rather on the operator at the other end. As I stated earlier; if we want to find out if he or she does QSL, or how, check their QRZ.com page. If they are worth their salt, they'll post it there. The same goes for those who do not require a SASE. They normally point it out on their QRZ.com page as well.

The electronic means of confirming a QSO is essentially made up of two different systems. One is LoTW by the ARRL and the other is eQSL.[39] Dave Morris, N5UP, first created QSLCard.com in 1998 and then eQSL in 2000, and remains to be the owner/webmaster. The eQSL structure has three membership levels available: Regular membership is free, Bronze and Silver both require an annual fee. *CQ Magazine* accepts eQSL for the CQ WAZ, CQ WPX, CQ DX, and CQ USA-CA awards. The ARRL does not recognize eQSL as a valid confirmation even though eQSL created an "Authenticity Guaranteed" program. Therefore, it is my belief, the eQSL company created its own set of awards, which more or less mirror the ARRL, e.g., Worked All States (WAS) and DX Century Club (DXCC). If you are satisfied with receiving an eQSL award and want to know more about the program, you may find them at https://www.eqsl.cc/qslcard/Index.cfm. However, if you want the *genuine* DXCC, or other ARRL awards, by using electronic confirmations, you will need to use LoTW.

Gary L. Drasch

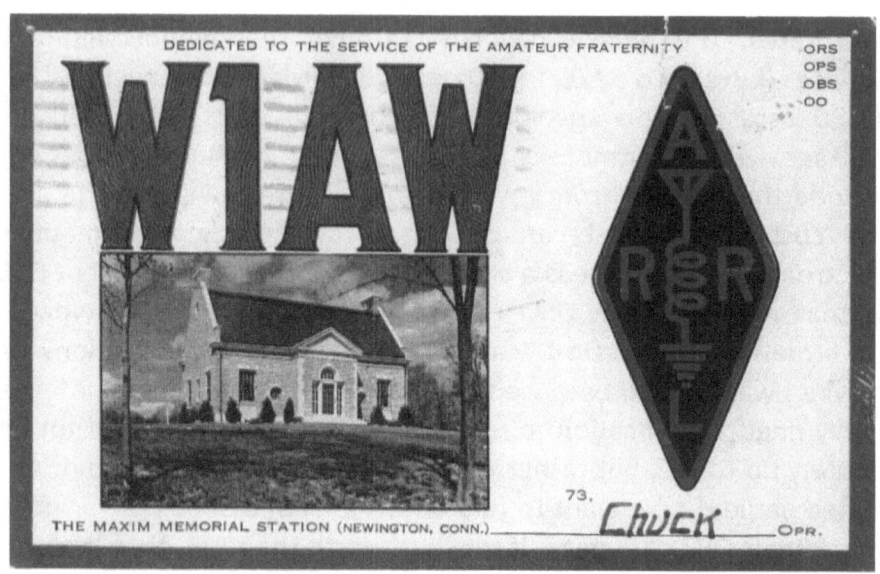

QSL card received at age 14, General Class license. Take note of the 3-cent stamp. (G. Drasch photo)

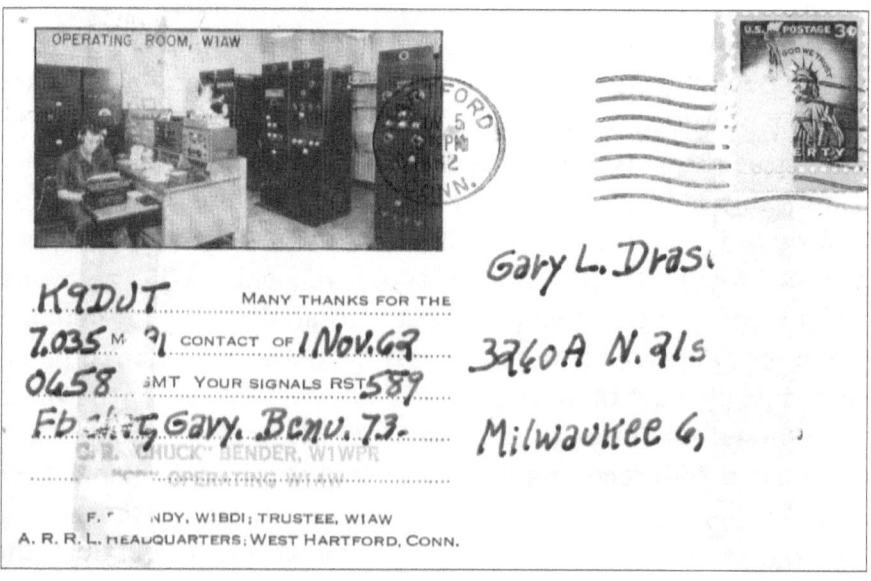

Back of Card. (G. Drasch photo)

Ham Radio is Alive and Well

QSL confirming a QSO with the DX-pedition on their way to Heard Island. (G. Drasch photo)

QSL confirming a QSO with a cargo ship on the Atlantic. (G. Drasch photo)

QSL confirming an aeronautical mobile QSO. (G. Drasch photo)

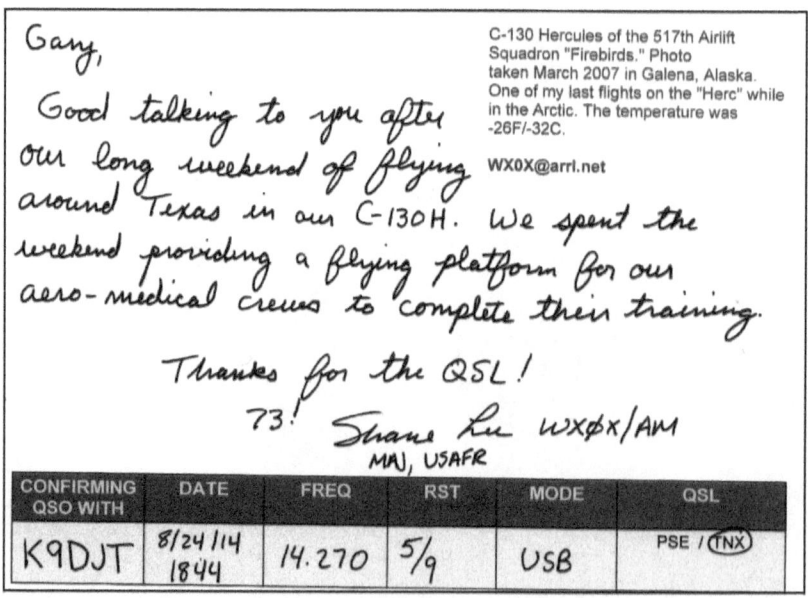

Back of QSL with a personal note. (G. Drasch photo)

Ham Radio is Alive and Well

QSL from King George Island. (G. Drasch photo)

QSL from Atka Bay. (G. Drasch photo)

Gary L. Drasch

The original idea of an electronic QSL at ARRL came about at a board meeting in the year 2000.[40] After various discussions on how the system might work, it was decided that it would be a *repository* of QSOs, rather than an exchange of emails. The Delta Division Director, Rick Roderick, K5UR, present-day ARRL President, spoke up and said, "So, this is like a logbook of the world?" hence the name. After loads of work by many people, LoTW became a reality. It was implemented on September 15, 2003. What I love about it is that it's FREE to use by any radio amateur, ARRL member or not. (Expect an application fee per award when using LoTW, which is no different than submitting QSL cards.) LoTW is accepted by *CQ magazine* for only the CQ WAZ (Worked All Zones) and CQ WPX (Worked All Prefixes) awards at this time. Negotiations to add others are in the works. If you are interested in using LoTW, you need to go to this website: https://lotw.arrl.org/lotw-help/ and make an application for a digitally signed *Trusted QSL* (TQSL) certificate. It is a little bit of a hassle, but simply follow the instructions, and you will be fine. I am happy with the system and have saved a ton of money using it. If you do implement LoTW, I highly recommend spending some additional time incorporating it into your logging program. Doing so automates the up and down loading process of QSOs (the only requirement is to insert the LoTW paths and passwords). You might also need to find the appropriate buttons in your logbook software and place them on the toolbar. It is well worth the time spent.

Once the integration is complete, in the simplest terms, this is how it works. Let's say you had several QSOs and you entered them into your logging program. At this stage, you need to highlight or tag the QSOs you want to upload to the ARRL LoTW server. Numerous logging programs have two buttons relating to LoTW, upload, and download. Press the upload button. The LoTW server identifies your TQSL, and a dialog box is displayed showing the progress of your upload. When it is done, click the *Finish* button and the dialog box closes and you are done. Those QSOs you uploaded have become QSO records in your LoTW account. What needs to happen afterward, is that the other stations, the ones you had the QSOs with, must upload their logbooks to LoTW too. Suppose a few of the stations you worked did upload their logs. LoTW then compares all your records, against all the records of other users, and finds there are matches. It then tags both sets of records as verified. Okay, let's say tomorrow night you decide to see if you have received any confirmations on those QSOs. (In LoTW, they are referred to as *verifications*.) You go to your logbook program and press the download button, and again a dialog box opens. This time there is a calendar displayed, and you select a date of how far back you want to go in time for a download. You should not select a date any earlier than the date you started uploading QSOs to the server. Press OK when done. At this juncture, the LoTW server again identifies your TQSL in your PC, begins downloading data, and tags all records in your logbook which had a match. Once those QSO confirmations (verifications) come through, you can start using the various tools in your logging program to sort and/or show your progress toward an award. You may alternatively go into your LoTW account on the ARRL website and view all your QSOs that have been verified and applied towards an award. LoTW is quickly becoming the standard for electronic confirmations; it is thoroughly supported, and uncomplicated to use once it is set up.

If the DX bug has bitten, and you are as serious about tracking your DX numbers and QSLing as I am, you should consider using *Club Log*.[41] It is an online tracking and statistics tool for DX-ers. The program was developed in 2007 by Michael Wells, G7VJR, and has approximately 244 Clubs, 56,267 users, and 89,158 call signs in its database. It works similarly to LoTW in that all the participants upload their logs to it, but then takes the results beyond the matching of QSOs. It allows users to create all sorts of charts and graphs showing their progression toward an award. Besides log matching, there is a Log Inspector which permits a user to test a call sign to see if it is valid or not. There are QSL charts that display matched QSOs along with OQRS (Online QSL Request Service) indicating buttons, which allow users to request a QSL through the service. Another feature is a DX cluster embedded within the site which compares a user log with recent spots of DX stations that are not in theirs. It shows the day and times they have been spotted, which then allows an operator the opportunity to chase those call signs the next day around the same time and frequency. The user will find different leagues they might want to join. These show how friends stack up against friends within a club, or how one club compares to others. Club Log shows the number of contacts and the percentage of contacts, he or she made using CW, phone, and digital each year. If you choose to try it, upload your complete log using the ADIF format (Amateur Data Interchange Format), and start poking around using the different buttons. You are not going to hurt or break the software, and you will discover how useful this tool is. All it takes to create an account is to sign up at https://secure.clublog.org/loginform.php. Recently, Club Log announced it has been accepted as the first Trusted Partner for LoTW. What that means, is we may take our LoTW TQSL certificate and upload it to Club Log. Once set up, we can mouse over to the lower-left corner of the Club Log menu bar and click on "Sync LoTW." And it does precisely that; it synchronizes LoTW with Club Log. We no longer need to upload our logbook to both servers, simply either or. I love it!

I cited OQRS[42] while explaining Club Log and I want to add that you may come across the same service when checking a DX station on QRZ.com or their website. I look at OQRS as a hybrid means of obtaining a QSL card because the request is done electronically, but the card is physically mailed. OQRS was created in March 2003 by DX-pedition leader Chris Sauvageot, DL5NAM, and then named and developed by Bernd Koch, DF3CB. They were both with the Sudan, ST0RY, DX-pedition at the time. OQRS is available in different ways and has become a conventional term. It is at your disposal on DX-peditions websites, QSL manager's sites, and Club Log. DX-peditions literally make thousands of contacts and in truth are not interested in receiving our cards. They want to accommodate us, the award chasers, with their QSL as quickly as possible. Using OQRS eliminates the cost, time, and need for the average ham to mail his QSL to the DX-pedition in order to receive theirs. This is the process. After choosing the appropriate website for your OQRS, you need to fill in your information into the online form, i.e., your call sign and mailing address of where you want your card sent. You then are given a couple of payment options. One might be no-charge, where the card is sent through the bureau (mucho slow). Another might be a minimum charge of $3-5, and another which allows you to make an additional donation to go towards the expenses incurred by the DX-pedition. You decide and pay the fee via your PayPal account. As long as the DX-pedition uses LoTW, you'll receive your LoTW verification more quickly and your QSL card so much faster.

It's time to start collecting those QSLs and Awards!

Gary L. Drasch

Chapter 7
Digital Modes

The digital modes weren't around 50 years ago as they are today. Okay, RTTY was, and if you were around then, you recall it as being radio Teletype®. The system used electromechanical teleprinters between two locations. The US Military embraced and widely used radio Teletype®[43] from the 1930s on into World War II after the Navy first successfully communicated between an aircraft and a ground station in 1922. I always thought those large, noisy, paper-guzzling, electromechanical teleprinters were cool, but then, I felt the same when I saw my first dot matrix printer. When the RTTY signal was heard on the ham bands, it was always those weird tones that randomly changed in frequency. The only hams using it then were gents who managed to obtain old surplus equipment, therefore the mode was not all that popular. Things changed in the 1980s when computers running emulation software started replacing teleprinters. RTTY became known as a digital mode, and by using a PC and a sound card interfaced with an HF transceiver, any ham was able to operate not only RTTY but other digital modes as well—without consuming any paper! PSK31 and RTTY had been the two renowned while others continued to sprout. Some of them were: PSK63, PSK125, MFSK, MT63, Hellschreiber, Packet, Pactor, Throb, Olivia, Contestia, Domino, Thor, and SSTV.

In 2001 Joe Taylor, K1JT, introduced a suite of weak signal[44] (WSJT) digital programs primarily for the VHF bands, but more specifically for moonbounce (EME) and meteor-scatter. The first was JT65, and then JT9; both being extremely sensitive and attractive to the EME guys, but also extremely slow, taking four minutes to complete a QSO. HF operators also started to adopt these modes on HF but

Gary L. Drasch

Fernschreiber T100 Siemens Teletype® machine. (Google for reuse photo)

the honeymoon soon fizzled because of the time involved. Soon, JTMS, MSK441, and JT6M began being used for meteor scatter and has since been replaced by MSK144. Not long after, Joe created another two modes for 6-meter weak signal *sporadic E,* called FT8 and FT4. These modes are so robust that they can decipher signals which are too weak to be heard by the human ear. FT8 has the capability of decoding signals close to 20dB below the noise floor and is considered a *game-changer* in that it only takes one minute to complete a QSO. Although FT8 was designed for VHF, 50 MHz, and above, It has become one of the most popular HF digital modes ever. You may download a copy from this link: https://sourceforge.net/projects/wsjt/. I need to note two other things which tie in with WSJT-X and those are JTAlert and PSKReport. Neither are required to operate FT8 but rather are

Ham Radio is Alive and Well

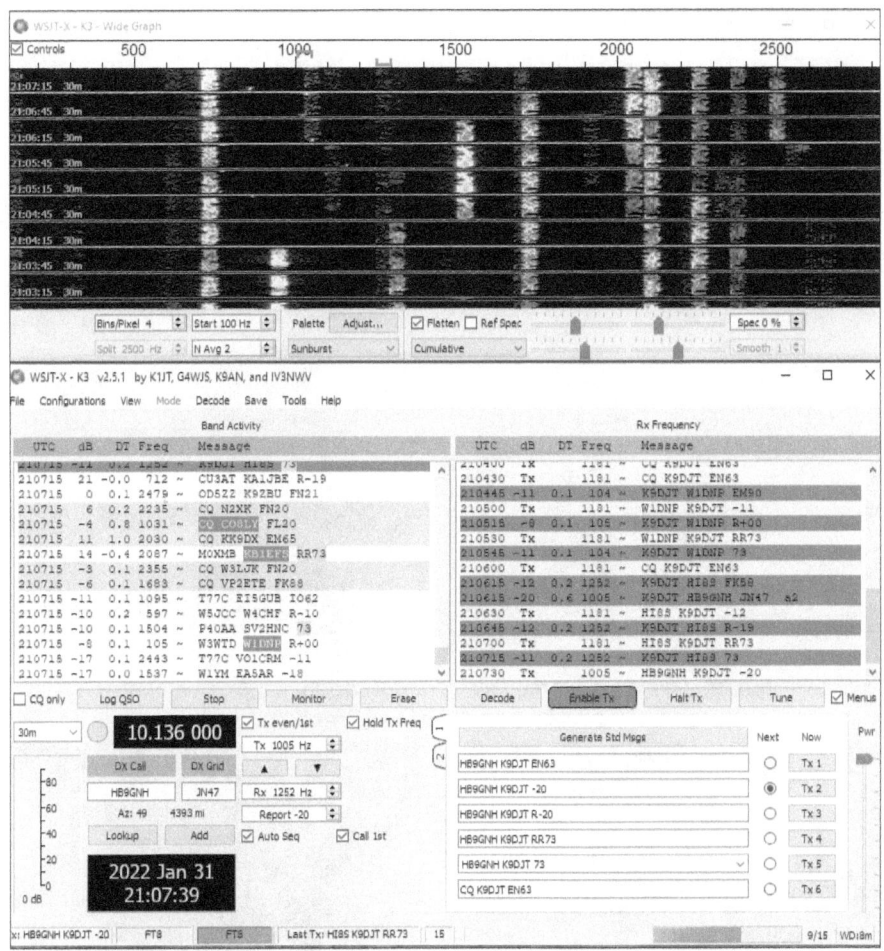

Screenshot of FT8 using the WSJT-X application. (G. Drasch photo)

niceties. *JTAlert* is an add-on to WSJT-X which compares your electronic logbook with what WSJT-X is receiving/decoding in real-time. For example, if you need a particular state for the Worked All States award, it can be set up to display that station in a certain color, and if you choose, announce it verbally through your PC speakers. The following link will take you right to it: https://hamapps.com/. The other cool component is a website called *PSKReporter* at https://pskreporter.info/pskmap.html. It displays the propagation of digital signals of your choice—even your own as you transmit. You will be amazed at who is all hearing you!

Another Taylor contribution is *WSPR*[45] (pronounced "whisper"), a program, and name, which stands for *Weak Signal Propagation Reporter*. WSPR collects signals and then generates a reception report which is uploaded to a database at WSPR.net.

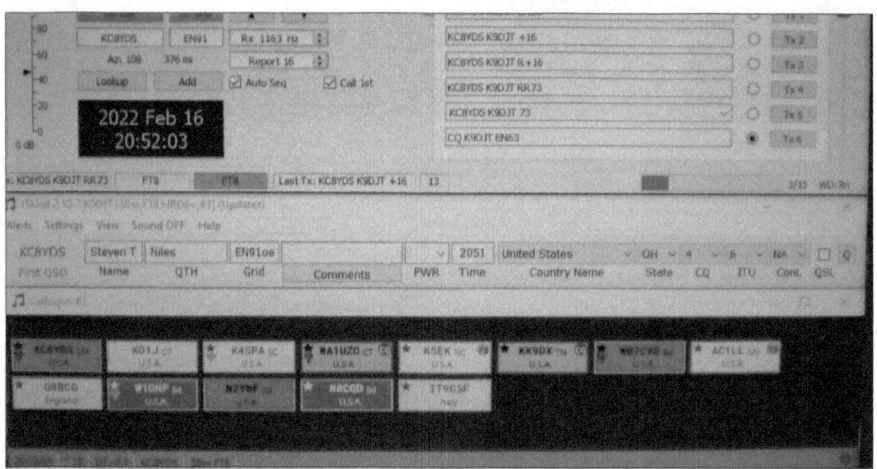

Screenshot of JTAlert below the WSJT-X application. (G. Drasch photo)

Screenshot of PSKReporter website. (G. Drasch photo)

As you may have already thought—is one monitor really enough?

Ham Radio is Alive and Well

A variant aspect of the digital modes is *Slow Scan TV*[46] (SSTV). It resembles a still picture sent on a FAX machine but is actually displayed on the PC monitor. SSTV, despite what its name implies, is considered a *digital mode* because it is a form of digital modulation. You'll find the SSTV signals at 7.033, 14.230, 21.340, and 28.680 MHz. MMSSTV, by Hamsoft, is the most popular software used. You can find it at https://hamsoft.ca/pages/mmsstv.php.

SSTV Screenshot of image received via the MMSSTV application.(G. Drasch photo)

A good list of digital modes; what they sound like, and the way they look on a PC waterfall display, can be found at http://www.hfradio.org.uk/html/digital_modes.html.

To become active using digital modes, you'll need to use the standard internal soundcard in your PC, and depending on your radio, you might need an additional soundcard between the radio

Gary L. Drasch

and your PC. My transceiver does not require one. I only need to run two audio cables, with 1/8" plugs at both ends, between it and the PC, audio-out of the radio to line-in on the PC, and audio-in of the radio from audio-out (speaker) of the PC. My Yaesu FT-2000 which I used to use at my cabin is different though. It required an external soundcard. The three popular brands of external soundcards, among others, are SignaLink, West Mountain Radio, and Timewave. Any of those, in addition to the normal CAT cable, will work well between the radio and the PC. The main purpose of the external soundcard is to handle the keying of the transmitter. My SignaLink connects to the computer using a USB cable, and then to the radio using a custom cable having an RJ-45 on one end and a 5-pin DIN on the other. The DIN plugs into the Packet port of the Yaesu. Nowadays, most of the newer radios have soundcards built-in. The only thing required is a USB cable between the radio and the PC. That one cable will handle the audio, keying of the transmitter, *and* CAT control. Pretty cool...

Besides the hardware, you must install a program on your PC, e.g., WSJT-X or MMSSTV. To utilize any of the legacy digital modes, i.e., RTTY, PSK31, etc., you'll require yet another software program(s). They are the interface between us mortals and the radio. My personal favorite is FLdigi, which does a variety of digital modes including CW. Yup, CW is a digital mode and can be decoded using this type of software (although the human brain greatly outperforms it). FLdigi performs well and is free. If you are only interested in PSK31 and PSK63, there is DigiPan; it too is a freeware program. Ham Radio Deluxe® (HRD) has a digital module, called Digital Master, as part of its logging program. Sad to say, it is only available as part of the logging package for which there is a charge.

The RTTY and PSK-31 software moderately resemble an email client, where we would type in our message and press send. As an alternative, a good typist can send a message in real-time; text shows up at the other end as he types. Although real-time sending is possible, the majority of operators do not operate that way. The

Ham Radio is Alive and Well

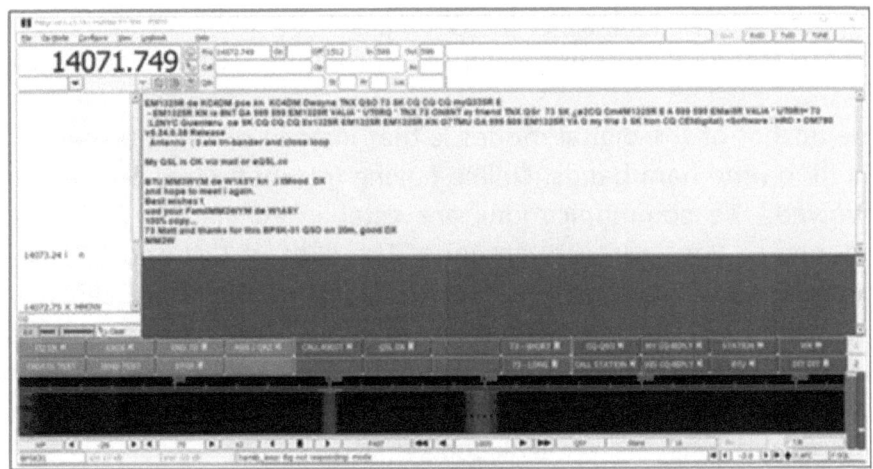

Screenshot of the FLdigi application operating PSK-31. (G. Drasch photo)

programs allow an operator to create various scripts, as many as 36 in some cases, which are sent by selecting and clicking an on-screen button. A script might look like this:

Thanks for coming back to me OM.
Your RSQ is 579 579.
The name here is Gary Gary.
QTH is Port Washington, WI Port Washington, WI.
Age is 74 74.
How copy? BTU
(His call sign) de K9DJT AR

The reply above is sent to the other operator, who was calling CQ, by the click of a software button. It is a common reply, which we might send to others who are calling CQ. Therefore, we saved it as a script under a software button of our choice. We may create other scripts relating to our station equipment, weather, and items we might enjoy talking about. Before we send the above, we want to send, "his call sign de our call sign." There is a log entry window, and dialog box, where we entered *his call sign* before we answered his CQ (the call sign remains there and is used during the entire QSO until we decide to log it). Everything can be done by clicking on various software buttons having an appropriate script.

The upshot of the digital modes is that instead of the Internet, it is via RF on the ham bands. Unlike having Internet wires and servers involved, the communications are established via RF, from your antenna to your contacts antenna. The likes of CW and SSB, it is instantaneous. Another plus for the digital modes is we can work DX using low power. What a fabulous tool for emergency communications! Two other digital modes requiring the same soundcard interface, and relating to EmComm (emergency communications) are WinLink and WinMor.

I first learned of WinLink from my friend Dr. Stan Kaplan, WB9RQR, who was the Emergency Coordinator (EC) of the Ozaukee Amateur Radio Emergency Services (OZARES) at the time. In fundamental terms, WinLink 2000 is a worldwide system supporting email via radio using non-commercial links to the Internet. It allows radio amateurs to send and receive email using a combination of standard Internet connectivity with radio-Internet connectivity in areas where standard Internet is unavailable. In any case, messages are composed and sent, or received and read, using a favorite email platform.

In the early days of email, I used a dial-up modem on my PC connected to a telephone line. I would create an email message in my program, and when done, I would click on a software button to dial up my email server. There would be a bunch of modem tones during the handshaking and it would then connect. My message was uploaded to my service provider and when done, if messages were waiting for me, they would be downloaded one at a time. WinLink works similarly, either through the radio or a working Internet connection. It is the same old process of data handshaking that takes time. Data transfer is slow; only 1200 baud for an HF radio connection. With that said, I can place an email directly onto the Internet via a radio. It is not as fast as what we are accustomed to, but consider how useful it is for emergency purposes. It is a standard component of the Amateur Radio Emergency Service (ARES) arsenal. If the Internet is disrupted, due to a local or regional emergency, ARES can respond with ham stations to send simple emails to victims' loved ones saying they are okay. Then add maritime mobile into the picture. There are yacht owners who have essentially obtained a ham license for the sole purpose of using email via WinLink.

With the goal of eliminating the need for additional hardware such as a Terminal Node Controller (TNC), Rick Muething, KN6KB, and Vic Poor, W5SMM developed the WinMor protocol along with the RMS Express program. RMS Express resembles an Outlook or Thunderbird email application. The WinMor protocol, introduced in 2008,[47] allows users to interface their radio with a PC using a soundcard that they might already use for other digital modes, e.g., PSK-31 or RTTY. No longer is PACTOR required to access the WinLink 2000 network. I used WinMor, preceding my smartphone, at my cabin to send and receive emails.

As far as soundcards go, I have used SignaLink for all of the digital modes. After it's connected, you need to install the RMS Express suite on your PC; WinMor is included. I found it painless to install, set up, and use. Once installed, you need to register your call sign, which creates your email account on the WinLink system. The RMS Express suite works with WinMor on the HF bands, which is the one I use, but it also works with the WinLink protocol on VHF. Unfortunately, VHF requires a TNC.

I struggle with the practicality of some Amateur Radio Emergency Services using VHF as the RF link. Using VHF for an RF communications link limits the usable distance required to find a node connected to the Internet. Additionally, there are not as many WinLink RF nodes on VHF as there are for WinMor on HF. Rather than only being able to communicate maybe one or two communities away; WinMor allows the user to connect to nodes

Gary L. Drasch

around all of North America. I have often sent an email using nodes in Texas, Canada, and Minnesota. Those types of distances are far more realistic to use for a major Internet failure and the reason I prefer exercising WinMor periodically.

The Amateur Radio Safety Foundation, Inc. (ARSFI) developed WinLink and supports it through an all-volunteer group. You can find the software and information at Amateur Radio Safety Foundation (ARSFI) and http://www.winlink.org. The RMS Express software suite, which includes WinMor, is downloadable and free. Although not mandatory, the ARSFI does appreciate a $25 per year charitable donation.

All of the digital modes are fantastic for *little pistols* (low power stations) operating out of an apartment with a vertical or some wire antenna hanging off their porch. Typically, 10 to 25 watts is adequate to get the job done. One more thing. I failed to mention a major benefit to using digital modes, and that is an operator does not need to hear well, or for that matter, at all. It's definitely a blessing for those hams who are struggling with a hearing impairment or loss.

Chapter 8

So Many Things to Do

Have you ever given any thought to the number of different facets there are in Amateur Radio? There is so much to learn and do that I find it hard to believe a person could take part in all of them during a lifetime. No matter one's age, physical condition, technical expertise, or operating abilities, there is stuff to do for all lifestyles. Hams in wheelchairs, those who are blind, and even the hearing impaired, can participate in this wonderful hobby.

Below are four lists of different things we can all do as radio amateurs. You will find several separately highlighted after the lists. My hope is that you'll find a couple that truly interests you.

Operating

- Hunt for DX.
- Contesting—try SO2R and/or hook up with a contest station.
- Rag Chew.
- Operate CW—operate in the ARRL Straight Key Night.
- Operate FT8/FT4, RTTY, and PSK. (digital modes).
- Operate portable as a QRP (low power) station.
- Do a 6-meter Grid-pedition.
- Make some EME (Earth-Moon-Earth) QSOs using WSJT.
- Chase an ARHAB (Amateur Radio High Altitude Balloon).
- Give Slow Scan TV (SSTV) a try using MMSSTV software.
- Access the ham radio satellites with a walkie-talkie.
- Work the ISS (International Space Station).
- Activate a park (POTA), a summit (SOTA), or an island (IOTA).
- Make contacts via Meteor Scatter using WSJT MSK144.
- Check into some informal nets—HF and FM.

- Become a net control operator.
- Participate in public service nets.
- Get involved with the—NTS (National Traffic System).

Community Involvement

- Join a local and/or regional radio amateur club(s).
- Become a VE (Volunteer Examiner).
- Take ham radio into a science class at your local high school or middle school.
- Volunteer as an ARRL manager or coordinator.
- Create and provide a presentation to a radio club.
- Become a QSL manager.
- Operate during Field Day or at a Special Event.
- Attend hamfests—maybe assist with running one through a club.
- Join ARES (Amateur Radio Emergency Service).
- Get involved with the Red Cross as a communications resource.
- Go on a DX-pedition.
- Help provide communications for parades, bike rides, walks, and runs.
- Get involved with the Amateur Radio Lighthouse Society, and/or active a lighthouse.
- Become an *Elmer* for a new ham.
- Teach licensing and/or CW classes.
- Do some Fox Hunting (Yup! The original Geocaching and Pokémon-go).

Technical

- Design and build your own antennas. (Especially wire ones!)
- Help others with their tower and antennas.
- Make your own cables and accessories.
- Plan and assemble your station. (Likely multiple times.)
- Build a kit. (They are making a comeback.)
- Study proper grounding techniques.

- Restore and/or collect boat anchors. (You know—the ones you always wanted as a kid.)
- Build a 12VDC power supply—design it or get plans from a book or article.
- Build a linear amplifier—there are numerous plans available.
- Construct a dummy load.
- Design and build an antenna tuner.
- Build a portable QRP station.
- Put a "Go-kit" together for emergency use.
- Install and host an FM or digital repeater.
- Set up and host a WinLink node.
- Collect and restore old keys and/or microphones.
- Set up a bench for experimenting. (Pick up ARRL's Hands-on Radio Experiments, 1 & 2, by Ward Silver, N0AX.)
- Refurbish old test equipment.
- Install a mobile HF and/or FM radio into your vehicle.

Are you interested in wallpapering your shack? There are numerous awards to chase—more than you can imagine. The list below is only a part of what I believe are the more popular ones. Every country represented by the IARU literally offers some form of an award for operating expertise. There is even a website hosted by K1BV dedicated to identifying awards worldwide at http://www.dxawards.com/.

Proficiency Awards:
- WAS (Work All States)
- Five-Band WAS
- The Triple Play Award (WAS using phone, CW, and digital modes).
- DXCC (DX Century Club).
- Five-band DXCC.
- DXCC Challenge.
- WAC (Worked All Continents).
- Five-band WAC.

- WAZ (Worked All Zones).
- Five-band WAZ.
- IOTA (Islands on the Air).
- SOTA (Summits on the Air).
- POTA (Parks on the Air).
- NPOTA (National Parks on the Air).
- WACA (Worked Antarctica Call signs Award).
- VUCC (VHF-UHF Century Club).
- FFMA (Fred Fish Memorial Award).
- County Hunting Award (*CQ Magazine's* USA-CA Award program).
- RCC (Rag Chewers' Club certificate - Society for the Preservation of Amateur Radio.
- Work towards a special one-time award, e.g., the Centennial Challenge.
- Create your own goal—as my friend Dave, KJ9I, by working as many stations as he can with "DX" as a suffix in their call sign.
- Collect QSL cards.
- Collect Certificates and/or QSL cards from Special Event Stations.

Did you find any fascinating enough to plow into? Are you able to see how some have the potential of becoming an obsession? With all the stuff listed, I am unable to understand why any active ham would be bored or need a job after retirement to keep busy.

Here is a list of activities I chose to highlight—what they are, and how they function. I'm only going to touch upon each one because there is so much specific information available elsewhere.

Rag Chewing

Many non-hams always ask me, "What do you talk about?" The answer is anything and everything. The conversation can range from radio equipment, antennas, DX-ing, Contesting, fishing, hunting, camping, family, work, history, geography, religion, politics, model railroading, piloting airplanes, to our daily aches and pains. Everything is open for discussion. It is no different from having coffee with a bunch of friends at some greasy spoon in the morning. We hams refer to it as *rag chewing*. Unlike some of the other things we diddle with, rag chewing has been around since the dawn of ham radio. You may have noticed there is even a certificate available for rag chewers.

I am bringing it up here for those who might be uncomfortable engaging. Having a conversation with a person we are meeting on the air for the first time should be pleasing. But it isn't necessarily so for all—especially for a bunch of us radio geeks. Every operator's personality is different so it is different for each of us. In many cases, technical and engineering-type people, which make up a large portion of ham radio operators, would much rather chase electrons around the shack (experimenting, building, configuring, and repairing) than talk with a human being. And that's okay; it's part of the hobby too.

An essential part of a good conversation is not as much about talking as it is about listening, and the way to get the other operator to talk is to ask questions. Common interests are the key to developing a conversation, and one of the easiest ways we can uncover those is by looking at the QRZ.com page of the other operator at the beginning of the QSO. Without delay, we can open up by saying, "I see on your QRZ page that you have a pair of Shelties. I only have one dog, a Golden Retriever, and I love him as family." Another opening might be, "So John, besides ham radio, what do you do for fun?" Or maybe, "I recently retired from the fire department after 35 years. Are you still working or are you retired too?" Those are the type of questions that will get things rolling. It's better yet if the other operator starts asking questions of us. The obvious commonality is ham radio, ranging from station equipment to antennas and how-to-do. I am always amazed at the number of knowledgeable people who make for interesting conversations.

During our QSOs, we need to be aware of the fact that our conversations are not private, and any number of people might be *reading the mail* (listening), even small children. For that reason, we should all be discreet regarding our subject matter. On the contrary, I hear operators who sound as if they just don't care.

Operate CW

If you are a person returning to this hobby after a 40 to 50-year hiatus, I know you know the code—the Morse code that is! As I, you have not forgotten it. You simply need some practice and polishing to regain your proficiency. There are many tools that can help you now compared to when we originally learned it. I cited some applications I use to help me practice code in the contesting chapter. As mentioned earlier, the Koch application on my smartphone is the one I use while driving alone. (It's much better than listening to the news nowadays.) Browse the Internet and you will find a few organizations dedicated to CW and keeping the art alive. All of them have one commonality; honing and improving the radio amateur's Morse code skills.

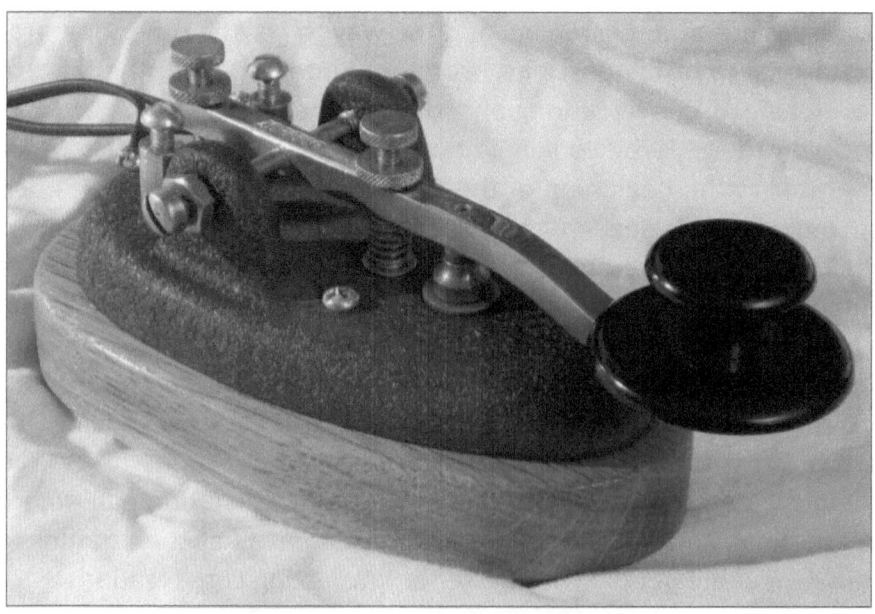

My Grimmer-Wilson straight key. (G. Drasch photo)

If you're interested in getting back on the air with CW and having some causal, slower-speed QSOs, I recommend you look into the Straight Key Century Club (SKCC). I joined them after a *Straight Key Night* on New Year's Eve. They have achievement awards AND a *monthly* Straight Key Night! Electronic keyers are not allowed. The whole idea is to keep the tradition alive by only using mechanical keys, i.e., straight keys, bugs, and side-swipers (cooties). Check them out at http://www.skccgroup.com/. Another one is The International Morse Preservation Society (FISTS). They have chapters in the Americas, Down Under, East Asia, and Europe promoting the preservation of the Morse code. Their informative website can be found at http://www.fists.org/. The last one I am going to comment on is the CW Operators Club (CWops) at http://www.cwops.org/. I have many DX-ing and contesting friends who are members of this group. In order to join, you will need to find three sponsors with who you have worked, using CW at least two times each in the last 12 months. One of the neat things is that they have a CW university. The information you need regarding membership, of not only CWops but the others too, can be found on their websites.

QRP

QRP wasn't much of a topic 50 years ago. Some present-day operators consider running less than 1000 watts as QRP. In retrospect, that would have put most of us in the QRP arena while we ran our 50 and 75-watt input transmitters back in the day. But things have changed. We did not have the technology 50 years ago to miniaturize radios as we do today. The recognized standard of QRP today is 10 watts input or 5 watts output—or less. All sorts of QRP stations are out there, from commercially manufactured to homebrew circuits stuffed into an Altoids tin. Although you will occasionally hear a few guys using SSB, QRP-ing is primarily done using CW or a digital mode.

Rick, NK9G, a friend of mine in our DX club, loves QRP-ing. He is better known as, "Mr. 88 and Sunny," because of his frequent

NK9G operating QRP portable at Sheridan Park in Cudahy, WI. (NK9G, R. Mc Gaver photo)

weather reports to us Wisconsinites while he winters in Arizona. It's not uncommon to hear Rick operating QRP portable during the winter months from Arizona. He has a separate operating position exclusively for QRP at home.

Could QRP be some form of addiction? I believe so. If you check out the ARRL website, you will find a couple of articles titled, "QRP: More Than a State of Mind," and another one, "Confessions of an Inveterate Milliwatter." That alone, says it all. Try it sometime. You will be amazed at what is possible with such little power.

Amateur Radio High Altitude Ballooning

ARHAB (Amateur Radio High Altitude Ballooning) is a means to study and educate enthusiasts of aerospace science using ham radio. The first ARHABs[48] took place in Germany around 1964, and then in Finland in 1967. To my knowledge, there was nothing similar taking place during those years in the United States, at least not locally.

This is how it flies (pun intended). A group of buffs, typically a club, decide they want to launch a HAB (High Altitude Balloon) to provide data to scientists on the ground. It will need to carry a payload of gear such as a radio with an *Automatic Packet Reporting System* (APRS), a Global Positioning System (GPS), weather station telemetry, a camera, and an FM repeater—anything the balloon is capable of carrying that can assist in providing data.

APRS is based on a ham radio digital communications information channel. It provides a ham who is operating mobile the ability to monitor any area for 10 to 30 minutes: messages, weather, bulletins, alerts, and a map of all the activity. The APRS data is typically transmitted on a single shared frequency in each country (in the United States it is 144.39-MHz), and it might be repeated through digipeaters (a digital repeater). APRS was developed by Bob Bruninga, WB4APR, in the 1980s.[49] The following website provides the details of how APRS operates: http://www.aprs.org/.

When APRS is used in association with the balloon, the telemetry might be stored onboard, but the important GPS coordinates are transmitted, then received and monitored by many hams tracking the movement and location of the balloon. This is done using a PC and special mapping software showing exactly where the balloon is at all times. It is not unusual to discover several hams using notebook PCs in their cars while they chase the balloon, or let us say, more importantly, its payload. Then there are those amateurs who follow the balloon on their home PC while they monitor the radio communications between the vehicles, all for the fun of it.

The latest radio balloon launches, missions, and data from past events can be found at http://arhab.org/. Enjoy!

Satellites and the International Space Station
It is hard to believe radio amateur satellites sprouted roots in December of 1961.[50] Right after the launches of the Explorer 1 satellite on January 31, 1958, a bunch of radio amateurs on the

west coast, began thinking of creating a satellite for ham radio. They later referred to themselves as project OSCAR (Orbiting Satellite Carrying Amateur Radio) with the idea of building and launching ham radio satellites. The group had conversations with the ARRL and the United States Air Force, which over time led to a piggyback ride of OSCAR 1 with the Discoverer 36 spacecraft. The craft was launched from Vandenberg Air Force Base in California, and by December 12, 1961, entered a low earth orbit. This had to be an exciting time. This 10-pound bird (satellite) which had been built in basements and garages was in orbit! OSCAR 1 was not capable of providing any type of two-way radio communications yet but rather carried a beacon transmitter, providing ionosphere propagation data and internal temperature telemetry of the bird. During its short existence of 22 days, over 570 hams in 28 countries, sent their observations to the OSCAR Project Team Data Collection Center. Although OSCAR 1 was lost as it burned on re-entry into the atmosphere, the event marked the beginning of ham radio into space. OSCAR 2 was carried aboard a Thor Agena B rocket on June 2, 1962. It was an improved version of OSCAR 1 and established a continuous improvement element to the amateur program. It was OSCAR 3 though which carried the first radio amateur linear transponder. It received signals near 146-MHz (the uplink) and retransmitted them near 144-MHz (the downlink). Around 1000 operators in 22 countries were heard using it during its 18 days of life. As of 2009, the OSCAR program has extended out to OSCAR 68 and is carrying an FM repeater.

On pages 1-20 in my 2010 edition of *The ARRL Satellite Handbook*, is a table of all the currently active amateur radio satellites. There are 19 listed, with one being the ISS, showing the frequency of the uplinks and downlinks along with the mode(s) of operation. If operating through satellites is of any interest to you, It is imperative you pick up a copy. Steve Ford, WB8IMY, is the author.

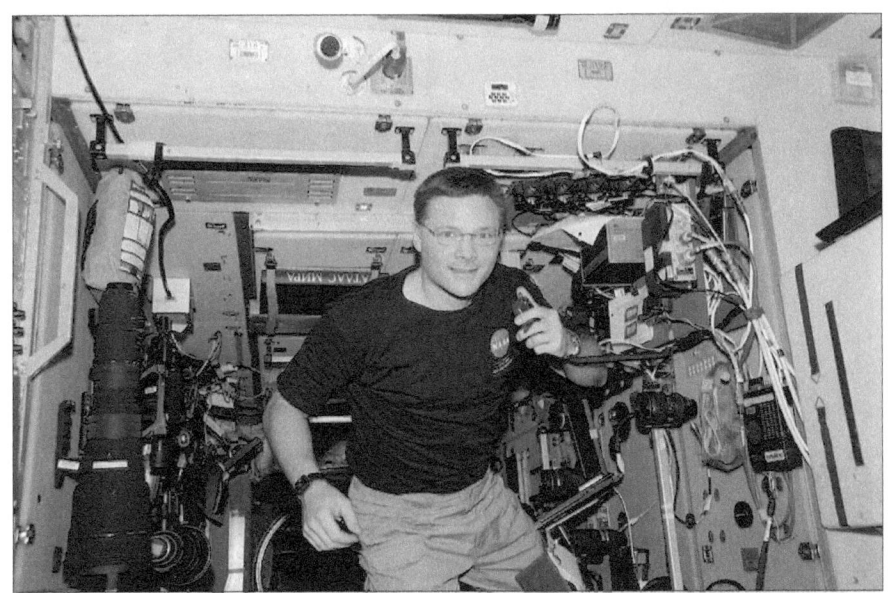

Doug "Wheels" Wheelock, KF5BOC, uses the ham radio system during ISS Expedition 24 and 25. (Google for reuse photo)

One of the cool things about operating through satellites is an operator requires only a Technician Class license, a dual-band walkie-talkie, and a budget antenna. It is probably the most inexpensive means of talking around the country on the ham bands. We might be in our backyard with a wide view of the sky, using a handheld radio with an antenna on a tripod, or using the same setup operating portable while looking out over a bluff somewhere. On the other end of the spectrum are larger radios and antennas using a multi-axle rotor, all interfaced with a PC, at a home QTH.

One of the key ingredients when operating through a satellite is to know where it is in relationship to our present location. Out of that need arose satellite-tracking software. The applications determine how high or low a bird is tracking our horizon while showing us what regions of the earth are in range for a QSO. Software such as this provides us with the next opportunity to intersect a path while displaying the satellite operating schedule, and its antenna orientation.

Gary L. Drasch

Antenna used for live student chat with International Space Station. (Google for reuse photo)

In addition, if we interface the PC with the antenna rotor, the software will control both the Azimuth and Elevation to track the bird as it travels. Connecting the PC to the radio also permits automatic frequency compensation to offset changes in the Doppler effect. Moreover, it is another reason to interface a personal computer with a radio!

If there is a downside to operating through satellites, it's the limited number of hams participating in the discipline. The same goes for working meteor scatter and EME. I've heard people say they have worked everyone who can be worked and lost interest. My answer is, "so what?" Go back and work them again...and again...and again! It isn't any different than those who check into a net every morning. Aren't they talking with the same guys over and over? Of course, they are. The difference with VHF and above is the science and mystique. There is so much more involved than just pressing a PTT button. And the best part; every day more and more operators are captivated and joining the fun.

A good satellite primer is downloadable from the ARRL website at http://www.arrl.org/files/file/Technology/tis/info/pdf/0004036.pdf (Steve Ford, WB8IMY, the former editor of *QST*.) By the same token, you might want to look into an organization called Amateur Radio in Space (AMSAT), which was created to continue the efforts started by Project OSCAR. Scope them out at http://www.amsat.org/.

Not only does the ISS carry a repeater onboard, but it is also possible to have a QSO with an astronaut. Dr. Owen K. Garriott, W5LFL, was the first astronaut to take a ham radio with him into space. It was November of 1983. He used a Motorola walkie-talkie held up to a 24-inch window of the Space Shuttle Columbia. That mission was the beginning of radio amateur activity in space, and hence The Space Amateur Radio Experiment (SAREX) was developed. Ham radio gear became a customary component on the ISS in 1997,[51] requiring more or less all astronauts to obtain an amateur radio license. One of the goals is to establish communications with classroom students worldwide. This is accomplished through ham radio clubs, which set up temporary, redundant, stations at various schools to ensure contact with the space station.

Not only is amateur radio connecting the space station with classrooms around the world, but it is also allowing SSTV images to be received by the rest of us. The Russians began transmitting images via SSTV on 145.8 MHz, using the PD120 mode, in 2008. A blog was established as a focal point for the SSTV activity at: http://ariss-sstv.blogspot.com/. You will find dates and times of planned transmissions posted there.

The ARISS picture gallery, at the link below, will give you an idea of what others are receiving via the SSTV downlink. https://www.spaceflightsoftware.com/ARISS_SSTV/index.php

Gary L. Drasch

SSTV picture copied from the ISS. (W9XT, G. Sutcliffe photo)

For more information on Amateur Radio on the International Space Station (ARISS) go to http://www.ariss.org/contact-the-iss.html and https://science.nasa.gov/science-news/science-at-nasa/2000/ast21aug_1/. I just find all this fascinating.

Earth-Moon-Earth

Have you ever considered bouncing a signal off the moon? Look around, and you will find a bunch of radio amateurs who do. I have a friend, Ken, W9GA, who is one of those. He is into anything and everything dealing with 50-MHz (6-meters) and above. Moonbounce is often referred to as EME, meaning Earth-Moon-Earth. It's a means of having a QSO with another operator by reflecting a signal off the surface of the moon. It's not a full-blown conversation where names and QTHs are exchanged. It resembles a DX contact, which is 5NN and bye. Instead of RSQ, the exchange is signal strength, in DB, and grid square. Q65, and JT65, are the digital modes of choice. These were created specifically for EME. As mentioned in the *Digital Modes* chapter; you'll need to download the K1JT program suite at https://physics.princeton.edu/pulsar/k1jt/wsjtx.html and have an appropriate audio interface.

It is common to see EME enthusiasts having a large antenna array made up of two to six, or even more, yagis stacked and phased. These are aimed at the moon with a dual-axis rotor, meaning Azimuth and Elevation capabilities. The antennas do not need to be high, but rather require enough room to be rotated and elevated at an appropriate angle. Without saying, there needs to be a clear view of the moon. Information, along with several interesting links, can be found at http://www.arrl.org/weak-signal-vhf-dx-meteor-scatter-eme-moonbounce. When there, take notice that these same weak signal techniques are applied to both EME and meteor-

2 meter EME antenna array. (W9XT, G. Sutcliffe photo)

scatter (which we will discuss next). As well, the following link may be of interest to the radio amateur astronomer: http://www.nitehawk.com/rasmit/.

Working Meteo Scatter

Do you remember being a kid and listening to shortwave radio? It was like *magic!* The ability to hear someone from the other side of the country, or better yet, in another country. I know my eyes were twinkling the entire time! Well, working *meteor scatter* reignited the wizardly within me—*hat, cape, and wand!!!*

Most hams are aware of *Sporadic E* propagation, i.e., metallic ion layers using the ionosphere, on 6- and 2-meters during the summer. The openings can be phenomenal at times. I've already worked Europe and Japan during Es. Now consider the winter months on those bands without propagation. Nothing. You might make a few contacts 25-30 miles away, maybe a little farther using FT8, but that's it. But wait! You haven't tried meteor scatter.

Set yourself up with a 6-meter antenna, preferably a beam, and a 100-watt radio. (Any 6-meter antenna will work.) Go to your WSJT-X software, then to mode, and select MSK144. Now select 50.260 MHz in the dropdown frequency box. You may have noticed that instead of a waterfall display, you now have a *Fast Graph*. MSK144 for meteor scatter works almost the same as FT8 but not exactly. A lot of information can be found in a tutorial written by Brad, AB4BA: *Simple Guide to Meteor Scatter MSK144*. Check it out at https://www.parkerradio.org/community/general/simple-guide-to-meteor-scatter-msk-144/ as to how to set up and operate. Also, make sure to register and sign in to the PingJockey (PJ) bulletin board at https://www.pingjockey.net/cgi-bin/pingtalk. Everyone on there is happy to answer questions in real-time as you operate. (There's even a chance yours truly will be on it!)

One of the most important things to remember while operating scatter is to have patience, patience, and still more patience. A

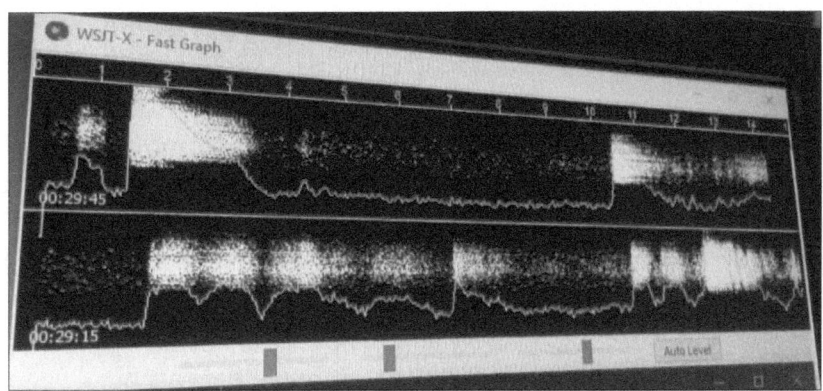

Screenshot of MSK144 Fast Graph showing duration of ionized meteor trails. (G. Drasch photo)

quick QSO is unlikely unless you're operating during a meteor shower. Plan on spending a few minutes to 15 minutes to complete a QSO during the winter doldrums. I once spent close to one hour to complete a 1,262-mile QSO on 6-meters! Think about it though—1,262 miles on 6-meters—*And NO propagation!!!*

So what does one do during that hour your software is managing the QSO? You communicate with the other guy on either PJ or Slack (https://vhf-chat.slack.com/). None of the QSO details such as signal reports or RRs are allowed to be discussed. That would disqualify the contact. Instead, you keyboard about other things like any other QSO, but on bulletin boards. It's definitely a different way of operating—but then again it is the 21st century!

Nets

A net is a group of hams who show up regularly on the same frequency, on specific days, at a particular time. There are local area nets, regional nets, and wide-area nets. If you have a license, you probably recall some of them from years ago. Collins equipment, model railroading, experimental aircraft, missionary, swap fests, astronomy, ARES, and weather are only a few of the special interest nets. They go on and on. A few nets in my locale are the Milwaukee Florida net, the NUT net, the Upper Peninsula net, the Mid-Cars net, the 68 Group net, the Buzzards net, and the Ozaukee Radio Club net. These are referred to as structured nets, meaning they all have a net control operator; a person who is in charge of running the net. He or she might do it for a week at a time or even months at a time. Some of the nets have a means of alternating net control operators regularly. At the beginning of a net, the control operator starts with a preamble explaining what the net is about and then requests check-ins, which he logs as a list. Once several people have checked in, he starts at the top of his list and calls the first check-in. He then asks either him or her to make a comment. This can be a specific interest to the group, such as a woodworking project, a model plane, to a problem he or she is having with a piece of gear. In some cases, each check-in comments about the weather at their location. Some might ask for a signal report from the group as to how strong his signal is and whether his audio is clean or not. It's a means of gathering a lot of data from a wide area. Rather than typing at a keyboard in an Internet forum, the participants are *actually* talking with one another live. Somewhere between the beginning and end, it becomes a round-table discussion. After the first check-in finishes with his comments, he turns it back to net control who calls and asks the second check-in to make their comments. This continues until he finishes the list. The net control operator might then ask for additional check-ins and add them to the list. At this stage, he (net control) starts at the top of the list again. Nets usually last about one hour, which means, depending on the number of participants, there can be two or three go-arounds. Some check-ins might comment during their first go-around that they are not going to be sticking around for the second. This is fine. There are many friendships created through nets. Some nets might have an annual picnic, breakfast, or lunch, at the same spot every year. This allows participants a chance to experience an *eyeball* (face-to-face) QSO. You'll find nets at different times of the day, and on all the bands, including 2-meters and 70-centimeters FM. There are tons of nets. So, how does a newbie, or an old-timer returning to the hobby, get involved? Ah, that is where the PC comes into play again. You can find all the structured nets within the ARRL directory at http://www.arrl.org/arrl-net-directory-search. The directory is searchable by NTS area nets, section nets, maritime nets, state nets, NTS regional nets, wide-coverage, and local nets. Then by US state, Canadian province, or US territory; net name, day of the week, frequency, and National Traffic Affiliation. The days, times, and frequency of each net are listed. No matter what your interest is, I am sure you will find a net to suit you.

I referred to *structured nets* a couple of times in the above. The reason is that there is another form of a net, which is more of a get-together than a net. It is common to find small groups of operators on 75 and 160-meters having a lengthy conversation. Although they ID (identify) themselves every ten minutes as required, they do not make use of a *net control operator*. In addition, they do not register themselves with the ARRL directory; therefore they are not listed. They talk as any bunch of geezers would in a garage or barn, and speak up whenever they have a word to say. It is all fine and good except no one can see one another's faces or mannerisms as one would when face-to-face. Consequently, frequent *doubling* goes on, i.e., two operators talk at the same time which royally screws things up. There was a time when I participated in such a group and decided to leave because I didn't find it fun to operate that way. Although the method of operation doesn't intrigue me, rightfully so, it does for others.

Of particular interest, under the subject of nets, is the National Traffic System (NTS) sponsored by the ARRL. The NTS is a network of radio amateurs who are skilled and organized in relaying formal traffic messages during times of emergency. The US and Canada are lucky to have such a backup system in place. You're welcome to join the group at http://www.arrl.org/nts.

Volunteer Examiners
I found it interesting to learn volunteer examiners date back to 1912,[52] when licensed radio amateurs administered the Amateur Second Class license. In the 1950s and 1960s, licensed radio

amateurs administered the Novice, Technician, and Conditional exams. (The Conditional license[53] had the same privileges as the General Class, but was for those living in a remote location and unable to travel to a Federal Building.) In 1984,[54] the FCC reverted back to using all volunteers. Thus, the VEC system (Volunteer Examination Coordinators) was created to administer the exams around the country, thereby eliminating the need for applicants to visit a federal building as I had in 1961. VEC organizations, which are responsible for the work of their certified VEs (Volunteer Examiners), are authorized by the FCC. The VEC is the liaison between the FCC and exam applicants. Hams who are interested in becoming a VE, need to hold a General Class license or above and be willing to volunteer their time to administer licensing exams through a VEC. New license, or the upgrade of an existing license, exams are given by a team of three or more VEs. Interested parties can find more information at http://www.arrl.org/become-an-arrl-ve.

This might be a good place to delve into a little history behind the restructuring of the Amateur Radio Service operator license.[55] In 1999, the FCC proposed the restructuring to reduce the number of exam elements and simplify the licensing requirements. It became official on April 15, 2000. Six years later, on December 15, 2006, the FCC disposed of the Morse code requirement. By doing so, they eliminated the disparity between the Technician and Technician Plus licenses. Furthermore, it opened the hobby up to many who had an interest in ham radio but were held back by the Morse code requirement. Today, there are only three licenses available. The Technician class, which is considered the entry-level license, requires a 35-question exam. The next is a General class requiring an applicant to hold a Technician class license and pass a different 35-question exam. And finally, the Extra class, requires a valid General or Advanced class license and satisfying a 50-question exam (the Advanced license had been grandfathered). None of the new licenses require a Morse code proficiency exam.

That is quite different from the 1951 structure which was Novice, Technician, Technician Plus (1990), Conditional, General, Advanced (1964), and Extra class (all licenses were valid for five years and renewable except for the Novice which was good for only one year and not renewable). In addition, all licenses required the passing of a Morse code proficiency exam. (The Novice class license did become renewable in 1978, and was grandfathered along with the Advanced class, into the 2006 changes.) It was in 1967 when the FCC introduced incentive licensing, where some band spectrum was taken away from General class licensees and divided up between the Advanced and Extra classes. The goal was to get Generals to increase their knowledge and skills to help prepare for the up-and-coming space age. The thing is, from the beginning, ham radio had always been a catalyst for many young people who later became engineers. It was a resource our government hoped to draw upon.

My written exam for the Novice license was 20-questions, and the General class was 50-questions, but I cannot recall how many questions there were in the Advanced or Extra class exam. One of my friends, Jim, K9QLP, helped me out. After some research, he found both the Advanced and Extra class license exams had a varied number of questions depending upon the license already held by the applicant. If he held a General class, he received credit for that exam. The same held true for the Extra class, if the applicant had passed both the General and Advanced class, he or she received credit for those and only needed to take another 50-question exam. If he or she did not have either of those licenses, the exam would have been 150-questions. Unlike now, anyone could apply for any class license without possessing a lower class. Along with those written exams came a Morse code test of 5-wpm for the Novice and Technician; 13-wpm for the Conditional, General, and Advanced; and 20-wpm for the Extra class. As the story goes, an Extra class licensee held the equivalent of a First-Class Commercial license during this period.

I have mixed emotions about the restructuring. On one hand, I can personally empathize with the amateur who needed to take a Morse code test for his license, where nowadays it is issued to any applicant who passes the written exam. Conversely, the restructuring opened the hobby up to many more people, which is one of the reasons amateur radio continues to grow. It is refreshing to see people carry on an interest in learning and using the Morse code, even though they are not required to do so. A big thumbs up to them!

Field Day

It does not matter if a person is a seasoned operator, who has been away from the hobby for a while, or if they are a newcomer, Field Day (FD) is one of the easiest ways to get involved with other radio amateurs. It takes place every year during the fourth full weekend of June. The ARRL originally organized the exercise in the United States in 1933.[56] According to their website, it has as many as 40,000[57] participants every year. FD is a public service and community outreach event demonstrating emergency capabilities while operating under less-than-optimal conditions. Many radio clubs and individual hams set up Emergency Operation Centers (EOCs) in public areas, e.g., parks, forests, beaches, and alike; using tents, trailers, and other temporary shelters. The power source is likely gas generators, or batteries being charged by solar panels. The operating goal, besides creating public awareness, is to make as many contacts as possible on the standard ham bands (WARC bands are not included) within 24 hours. Many people come and go while others might stay and camp overnight. Although FD is considered an emergency simulation and not a contest, scores are kept and published in *QST*. In my opinion, if we are keeping score, how can anyone say it is not a contest? The fact of the matter is, it's fun; a means of meeting other radio amateurs and becoming an active shareholder in ham radio. Find your local radio club and ask to join them next June. I am confident you will be welcome. It will allow you to see how the new technology intertwines with the old. As far as that goes, FD is a perfect way for *anyone* who has never been a ham to get on the air and try it for the first time. Many clubs have an operator-assisted GOTA (Get on the Air) station specifically set up for unlicensed persons to try.

Because of all the food and drink, you should plan to arrive with a good appetite. It is rumored some clubs consider FD to be more about eating than operating!

Amateur Radio Emergency Services

The ARRL *Amateur Radio Emergency Communications*[58] (AREC) unit began responding to disasters across the country in the 1930s. The AREC has since been replaced by *ARES* (Amateur Radio Emergency Services). It is made up of licensed radio amateurs who are willing to volunteer their time and communications equipment during a disaster. Those who are serious register with their local ARES group, listing their qualifications and equipment. Although ARES is sponsored by the ARRL, any licensed ham is welcome to join and is not required to be a member. Training may be required depending on the type of emergency services they provide. You'll need to inquire at the local level. Owning emergency power-generating equipment is a plus, but not required to be a member.

My county has an active ARES group named OZARES (the "OZ" stands for Ozaukee county). They host two open repeaters, a VHF and UHF; have monthly meetings and training sessions. Information about ARES can be found at http://www.arrl.org/ares. In Wisconsin, all ARES members are also members of *RACES* (Radio Amateur Civil Emergency Service). Whereas ARES is sponsored by the ARRL, RACES base sponsorship is provided by the US Government: Federal Emergency Management Agency (FEMA). Another government program, which is supported by ARES, is called SKYWARN. It was developed by NOAA's National Weather Service (NWS); the purpose is to obtain vital weather information, and visual confirmation of such, through radio amateurs in the field.

Gary L. Drasch

W9XT operating CW at Ozaukee Radio Club Field Day (G. Drasch photo)

Many Red-Cross volunteers have now become VHF-UHF operators by acquiring a Technician Class license after the code requirement was dropped. That dovetails in so well with ARES because most communications are performed using VHF-UHF radios. A shack is not necessary. Hams have set up VHF-UHF radios within their homes, in the basement, living room, kitchen, and bedroom. Some may have placed a dual-band antenna outside to communicate through a repeater located a significant distance from their QTH. I have beams on both VHF and UHF but in addition, I have a couple of homebrew verticals in my attic as backups. They both work exceptionally well for the local *machines* (repeaters). An operator might even get by with a handheld walkie-talkie depending on how close he is to a machine. If he uses a handheld, nothing is saying he cannot hook it up to an antenna in the attic or outside. I'm able to easily *hit* (contact) our club repeater, fifteen miles away with full quieting (no frying noise), using my 220-MHz walkie-talkie in my basement and an antenna in the attic. These bands are important for local emergencies where phone systems, cell phones, and

Internet services become inoperative, but what about the big picture? How would we communicate around the state or country? The Red Cross brings in satellite dishes to access the Internet, but what if their satellite equipment malfunctions? What happens if there is a catastrophic failure of the Internet? Are you thinking this is where HF comes into play? You're right. The Red Cross has already started implementing a digital protocol to be used on 30-meters for unassisted communications. The caveat is that a General Class licensee is required to oversee the operation of a radio at that frequency. I have confidence the Red Cross has at least one General Class licensee on-site during mock operations, as OZARES does. During a TRUE emergency, where life is in danger, no one is going to care about a license or what frequency is being used. What really counts is that the radio operator knows what he or she is doing.

FM Repeaters and Internet Linking

Thinking back, this is what I thought was left of the hobby until Lyle, WE9R, straightened me out. Although the concept of repeaters started in the 1930s,[59] it was not until the 1950s that AM repeaters started surfacing, but it was the 1970s when the FM repeater took hold and revolutionized the VHF and UHF, amateur bands. An abundance of surplus FM commercial equipment happened to surface in the 1960s due to a growing demand for newer technology. Hams discovered all this commercial gear was serviceable and able to operate on amateur frequencies, specifically the 6-meter, 2-meter, and 70-cm bands. This was not only mobile radios but walkie-talkies and repeaters as well. During this time, radio amateurs modified these commercial radios to work on their ham bands. After all, they had already been homebrewing most of their HF equipment—why not VHF and UHF? My grandfather would have been one of those guys if it had not been for the code requirement back then. He would have done quite well modifying and using that stuff. And he would have been delighted to talk with his friends across town through a repeater using a handheld or mobile radio.

Gary L. Drasch

I had a friend in the Ozaukee Radio Club, Nels, WA9JOB (SK), who primarily operated using the club repeaters, even though he held an Extra Class license and owned an HF transceiver. Why? Because FM and his friends on those bands were what was important to him. It is what he enjoyed. By the licensing numbers, it appears many people enjoy doing as Nels had. 50 percent of our licensees in the United States hold a Technician class license which limits them to two small portions of the 10-meter band, and then 6-meters and above. They also have some CW privileges on 15 and 40-meters but need to learn the Morse code to use them. I believe that most Technicians primarily use the 2-meter and 70-cm bands through FM repeaters, while some might use a 220 MHz machine if available.

I referenced EchoLink and IRLP earlier—the ability to connect to an FM repeater through a PC using *VoIP* (Voice over Internet Protocol). It all began in 1997 with the *Internet Radio Linking Project*[60] (IRLP). It continues today and runs on the Linux platform. By 2004, there were over 1200 repeaters and simplex stations using it. This opened the door to the development of another linking protocol called EchoLink. Because of the popularity of MS Windows, Jonathan Taylor, K1RFD, created the *EchoLink* program and wrote a book, *VoIP—Internet Linking for Radio Amateurs*, an ARRL publication. There are pros and cons to both systems. The true computer geek seems to prefer IRLP while the common Joe sticks with MS Windows and EchoLink. No right or wrong—just different.

Nels, WA9JOB (SK), periodically traveled to visit family around the country and, as the majority of people do, carried a smartphone with him. Nels not only used the standard features of his phone but additionally used it as a pseudo handheld radio by having installed an EchoLink application on it. It was common to hear him on a local UHF repeater talking with his pals while traveling. As I said, EchoLink and IRLP both use VoIP which is the same thing cable companies use to bring telephone service into our homes. As one PC can communicate with another PC on the Internet via Skype or ZOOM, EchoLink and IRLP can do the same using a PC, having a microphone and speaker, to communicate with another PC that is connected to a radio or a repeater. With EchoLink, it is done through an EchoLink software application installed on the PC. Hams can connect via a local repeater to a remote repeater of choice, or a PC to a repeater of choice or reverse, via the Internet. All of it is done using the keypad on the microphone of our FM transceiver. An EchoLink RF site, i.e., a link, can be created and used by any amateur licensed operator. I had done this once upon a time. I dedicated a PC and a UHF radio which I interconnected with a RigBlaster soundcard. The PC was running the EchoLink software and had an Internet connection. The radio, which was connected to the PC through the RigBlaster, transmitted and received using the same frequencies as any other VHF-UHF radio accessing a repeater. In my case, it was a UHF machine in Plymouth, Wisconsin. Hams selecting my node in the EchoLink lookup table could connect to my friend's repeater via the Internet. Plus the reverse was possible. Hams were able to access my friend's repeater using a handheld or mobile radio, and then connect to a different repeater anywhere on the planet by selecting it via their microphone keypad.

The Japan Amateur Radio League (JARL) developed a digital voice and data protocol in the late 1990s which is referred to as *D-STAR*[61] (Digital Smart Technologies for Amateur Radio). The main advantage of D-STAR is that it uses less bandwidth. It uses less bandwidth than SSB, AM, or FM. Icom, Kenwood, and FlexRadio Systems manufacture VHF, UHF, and HF products that are compatible with the D-STAR protocol. Unlike EchoLink and IRLP, all the D-STAR repeaters can find specific radios on the system through its Internet connection. There is no need to call up any machine first. Instead, the system finds the amateur we are looking for through the D-STAR network. Users of D-STAR can communicate via digital voice and short data messages, exchange photos using an IP camera, and access the Internet using a PC radio combination. Sadly, a D-STAR repeater cannot be accessed via an FM transceiver; it only works with a D-STAR protocol-type radio. D-STAR radios also include a GPS and support APRS.

There is a thing going on within the VHF/UHF digital modes in amateur radio which is not much different from the historic battle between the Beta Max and VHS video formats. Yaesu introduced their digital version called *System Fusion* not long after D-STAR. Sometimes System Fusion is referred to as C4FM, which stands for *Continuous Four Level Frequency Modulation*. From what I understand, both systems are similar except System Fusion has Automatic Mode Select (AMS™). The Yaesu digital repeaters and digital transceivers can accommodate FM as well as digital, and automatically selects the correct mode based on what is received. System Fusion supports APRS and has a GPS as well.

Yet another mode is *Digital Mobile Radio* (DMR). The European Telecommunications Standards Institute (ETSI) created this open digital standard to be used around the world in commercial radio products. It requires a programmable *code plugin* to operate with other radios in a group—great for a local club but eliminates your ability to move from city to city like D-Star or System Fusion. Hardware manufacturers include such companies as Ailunce, AnyTone, BTECH, Baofeng, BridgeCom Systems, Radioddity, and TYT. Both repeaters and hotspots are available.

Being no different than when SSB originally emerged, these new modes also have skeptics. Because bandwidth is such a concern in all electronic communications today, I cannot help but believe digital radios are the future. After all, it is already taking place on the commercial side of the two-way radio business.

Islands on the Air

Islands on the Air (IOTA) is an awards program run through the Radio Society of Great Britain (RSGB): the British equivalent of the ARRL. They took the program over from Geoff Watts[62] in 1985. Geoff was a shortwave listener and came up with the idea in 1964. The purpose was to encourage amateur radio operators to make contacts with other hams located on islands.

IOTA consists of 1200 groups of ocean islands, with different qualifying islands in each. The qualifying islands are referred to as *counters*. The goal for an island *chaser* is to confirm at least one counter in as many of these groups as he can. That sets him up for the possibility to earn up to 18 different certificates, graded by difficulty. The activator, the one who sets up and operates from an island, qualifies for an award as well.

A fellow member of my DX club, Wayne Long, K9YNF (SK), always spoke of what he had accomplished in the IOTA program. Wayne reported islands he had worked during the time the rest of us were reporting DXCC slots. He always referred to an IOTA award as being prestigious and rightly so. He had worked 710 islands and was proud to be on the *IOTA Honor Roll*. (Wayne was also a *DXCC Honor Roll* member.) Rest in peace, my friend...

You will find IOTA has become one of the fastest-growing awards programs around. Obtain all the information you need at https://www.iota-world.org/.

Summits on the Air

It was March 2, 2002, when *Summits on the Air* [63] (SOTA) became a reality. England and Wales adopted the program first, followed by Scotland. The idea of creating an awards program was the brainchild of John Linford, G3WGV, who came across Richard Newstead, G3CW. Richard was running a website called *European Adventure Radio* at the time. It was decided SOTA was to be an awards program and not a club or society. That decision made it possible for all radio amateurs and *shortwave listeners* (SWL) to participate without the need to be a member of any group. There are awards for *activators* of summits and those *chasers* operating from their homes. This means it is not only for mountaineers but rather for radio amateurs worldwide. Hundreds of countries are participating, with each one having its own association. Each association decides on what summits should be recognized by SOTA within its own area. Even a local hilltop, e.g., 984 feet high, might be considered a summit. A list of all the international summits, including all those in the USA, can be found at http://www.sota.org.uk/. We can click on our call sign area and then the state of interest to see what qualifies in that locale. The scoring for both the activators and chasers is based on the height of the summit worked or activated. Various scores notably earn different award certificates. The ultimate awards are the **Mountain Goat and Shack Sloth** trophies. The Honor Roll for both **Activators and Chasers** may be found on the SOTA website.

Worked Antarctic Call signs Award

For me, it all started after reading the book *Endurance,* an Antarctica expedition led by Ernest Shackleton. For whatever reason, Antarctica has always fascinated me. They now have Internet and TV via satellite, but nevertheless, ham radio continues to be popular with the various countries doing research there. All of their ham radio activity is the very reason for the Antarctica awards program.

The *Worldwide Antarctic Program*[64] (*WAP*) was founded in 1979 and is dedicated to ham radio Antarctic chasers. Note their website at http://www.waponline.it/. When looking at the WAP awards, the Worldwide Antarctic Awards Gallery, you will find awards available from Argentina, Australia, Italy, Russia, Ukraine, and Poland. Additionally, there are awards for sub-arctic areas and even a Polar Ship Award. I am proud to be a recipient of the *Worked Antarctic Call signs Award* (WACA) by submitting proof of working at least 10 different call signs in Antarctica. Since receiving the award, I have worked and confirmed an additional nine. I find it alluring—actually magical—to be able to communicate with a person, antenna-to-antenna, at the opposite end of the globe. Although many arctic operators use LoTW, I always enjoy exchanging QSL cards with them.

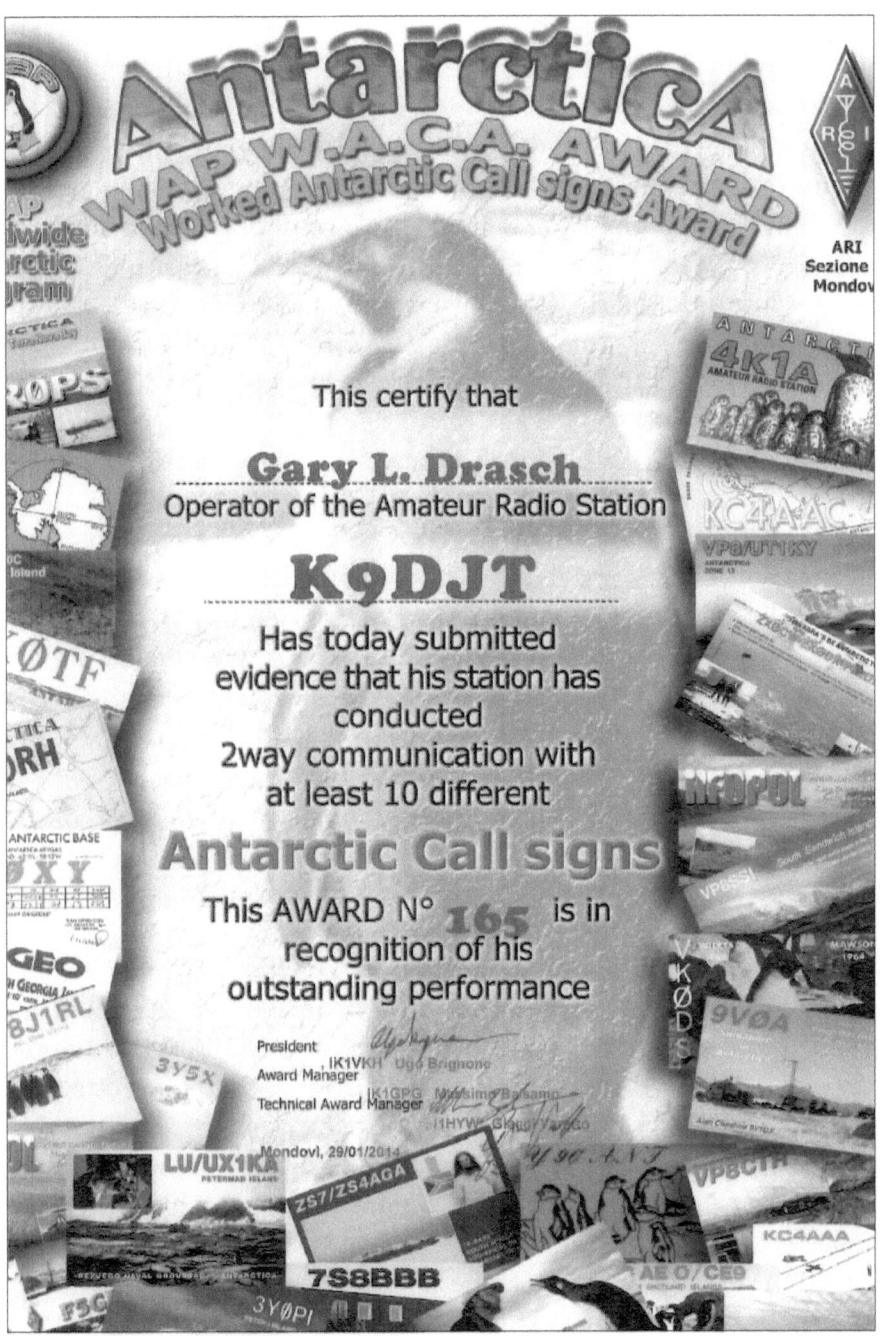

Worked Antarctic Call signs Award (G. Drasch photo)

Gary L. Drasch

VHF-UHF Century Club

Are you ready to hunt some *grid squares*? The place to do it is on 6-meters and above. My past 6-meter experience amounted to the homebrewing of a crystal-controlled AM transmitter—the goal of talking with friends in my local club. At the time, grid squares did not exist. Grid-squares are a part of the *Maidenhead Locator System*[65] describing our general location anywhere on earth. The name came about after a 1980 European VHF meeting that was held in a town outside of London called Maidenhead. This was definitely a new one on me!

Grid-squares are useful on all bands, especially when working a maritime mobile (boat/ship) in the middle of some ocean. When a maritime mobile station provides its grid square, we can locate the ship's approximate location on the ocean. A grid square is 1° latitude by 2° longitude and measures approximately 70 x 100 miles in the continental United States. It has two letters, which are the degree-field, and two numbers called the degree-square. For example, my home QTH grid square is EN63, and my cabin is EN55. To obtain a better resolution, there are sub-squares assigned. A sub-square is 2.5 minutes latitude by 5 minutes longitude, which equates to 2.5 x 5 miles. In my case, they are bj and sn respectively. Therefore, they are written as EN63bj and EN55sn.

The VUCC award was created in January 1983[66] by the ARRL. It can be obtained by those operators who have worked 100 different grid-squares on 50-MHz, 144-MHz, 220-MHz, 432-MHz, and above combined. Satellite contacts also count. You will find 6-meters to be an absolute blast when it's open during the summer. Up to now, I have worked 606 grid-squares and 61 entities (countries), but compared to Ken, W9GA, who I spoke of earlier, this is nothing. Besides being into EME, he is an avid VHF DX-er. Ken has an addiction to VHF. He is also a terrific contester who primarily focuses on VHF contests. Thus far, Ken has confirmed 136 entities on 6-meters, 34 on 2-meters, and a total of 712 grid squares.

Ham Radio is Alive and Well

Janice's "Little Red Rover." KA9VVQ Grid-pedition (B. Richardson photo)

Fred Fish Memorial Award—FFMA

Fred Fish, W5FF (SK), was the first radio amateur to have worked and confirmed all *488 grid squares* in the 48 contiguous States on 6-meters. The FFMA was created in his honor and is bestowed upon those who can match his achievement. It is very prestigious in that only 23 have been awarded since its conception in 2008. I, and many of my friends, keep pecking away at it. Not all grids are occupied by hams which is why it's so challenging. To help alleviate the situation, *Grid-peditions* arose. This is where one or two people decide to activate a *most wanted grid* by driving to it. They'll set up an antenna, and a station, and operate out of a vehicle, camper, or tent. Most stay a couple of days and move on to another rare one. As you can imagine, the award program encourages 6-meter usage and helps activate unoccupied grids. Use this link to learn all about the rules, rankings, and even the rare grids: http://www.arrl.org/ffma. Come join the fun!

Worked All Counties - USA

I was proud to have received my first WAS certificate at age 14 in November of 1962. I managed to do it again in each of the three modes during my retirement. But compare that to working All Counties—that's right—COUNTIES! There are 3077 of them in the United States.

The award was originally offered in the early 1960s[67] but was first realized and awarded to K9EAB (SK) on August 15, 1965.[68] To date, 1268 operators have captured it, with CU3AA being the latest recipient on July 27, 2019. Incredible! The program is one of many run by *CQ Magazine*. The official rules can be found at http://www.cq-amateur-radio.com/cq_awards/cq_usa_ca_awards/cq_usa_ca_awards.html. There are seven classes of awards offered: USA-500, -1000, -1500, -2000, -2500, -3000, and the Pinnacle USA-3077. There are several County Hunter nets; one of the more popular is on 14.336 MHz. They even have their own "spotting" network at ch.w6rk.com. It's not only a program for those who live in the United States. I have often met European operators on the air working towards the award too. Go ahead—tell me why you're bored!

Chapter 9

From Shack Heaters to Now

For some hams, collecting and refurbishing boat anchors is as much of the hobby as it is getting on the air. I recall the rigs from the mid-1950s and 60s such as Collins, R.L. Drake, Hallicrafters, Heathkit, National, Hammarlund, E.F. Johnson, Eico, Knight-kit, Globe, and RME; many have survived and are around in a ham's shack or basement somewhere. It's common to hear many of them on 40, 75, and 160-meter nets. If you were to participate in Straight Key Night on New Year's Eve or join in with the Straight Key Century Club, you will discover you are listening to some of the old crystal-controlled DX-40s and Globe Chiefs, and being told the receiver is an SX-100 or NC-88. It was in the mid-1960s[73] when SSB transceivers, namely the Collins KWM-2, Hallicrafters SR-150, and Heathkit SB-100 transformed and simplified the well-acknowledged amateur radio stations of yesterday. These new radios combined the receiver and transmitter into one package, sharing the same on-off switch, tuning knob, power supply, and many other common circuits between the two. Part of the reason AM was replaced by SSB was due to these sleek transceivers. It wasn't long until Swan Engineering recognized what was going on in the market and decided to get in on the SSB transceiver revolution. They introduced the Swan-350 and -500 around 1965 which perhaps became the best-selling radios during that time. This was a completely new market! Kenwood, known as Trio Kenwood, opened an office in the United States right about this time. Yaesu, which was founded as Yaesu Musen, also arrived in 1965.[69] Although the first Kenwood and Yaesu radios were all tubes, they both became stiff competition for the old pillars, in particular, Collins, Hallicrafters, and Drake. Around 1963[70] an American company, by the name of Sideband Engineers (SBE), introduced the SB-33. It was a hybrid transceiver that was completely solid-state except for three tubes in the power amplifier. Later, SBE came to be known as Robyn Manufacturing. Shortly after all this, the Japanese manufacturers followed suit and started importing hybrid transceivers while the rest of the real competition in the United States persevered in manufacturing 100% tube shack warmers. The Japanese became the dominant player in 1979,[71] which happened to be the same year Icom entered the scene. Ultimately, by 1984[73] the stateside manufacturers decided to leave our beloved marketplace, while several continued with their commercial business.

I uncovered an interesting and ironic tidbit on the Icom website during my research. It was about an interview *CQ Magazine* had conducted with the founder of Icom, Mr. Inoue. The meeting took place in 2001 and the writer was Sam Vigil, WA6NGH; the title: CQ Interviews;[72] *"Mr. ICOM"* – Tokuzo Inoue, JA3FA. I could not help but smile after reading the following excerpt of the interview: Mr. Inoue's business philosophy, from the very start of his company, has always been "Technology first, the money will follow." He was greatly influenced by this philosophy when meeting the late Arthur Collins (of Collins Radio), who gave him this advice: "No matter what, keep perfecting your technology. If you perfect your technology and make good products, you will always get business. Forget about unnecessary things and strive to exist by your technology." So, what do you think? Although it was fantastic advice, I wonder if Arthur had ever looked back and kicked himself saying, "Why did I ever tell him that?"

I have always admired Collins' equipment, from styling to the military grade construction. My friend Bill, W9MXQ, refers to Collins as having a desk presence. I had an opportunity to use a Collins KWM-2A at Field Day in 1964 and at the time was thinking of how much I wanted one. I always felt that way about their S-Line as well. I have told friends, "Now that I can afford it, I'm not interested in

Ham Radio is Alive and Well

Collins S-Line, 3-Series station (W9MXQ, W. Shadid photo)

Collins Gold Dust Twin station. (W9MXQ, W. Shadid photo)

Drake C-Line Station. (W9MXQ, W. Shadid photo)

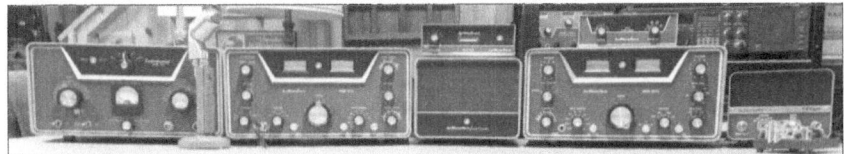

Hallicrafters SX-117, HT-44, HT-45 Station. (W9MXQ, W. Shadid photo)

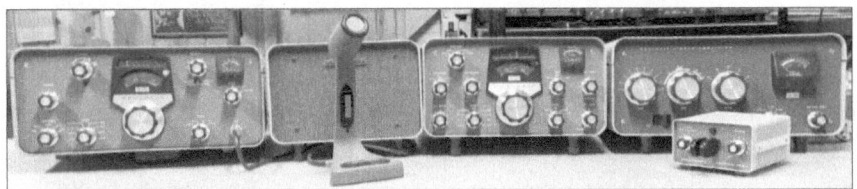

Heathkit SB-303, SB-401 Station. (W9MXQ, W. Shadid photo)

owning them." On the other hand, I am not going to completely rule it out. Without question, I go along having fond memories of the Gold Dust Twins from when I was twelve years old and spoke into a microphone for the very first time at the shack of my grandfather's neighbor. At present, the only vintage gear I have is my Grimmer-Wilson straight key and Champion Vibroplex bug and wishing I had held onto my HQ-170C receiver, Viking Valiant transmitter, and Astatic D104 microphone.

Recently, I asked my friend Bill, AC9JV, what drew him into vintage gear. Bill is an EE (Electrical Engineer), about sixty-five years old, who entered ham radio about six years ago. He said, "It's just cool, and it's something you can still work on." Bill is also one of those new hams learning the Morse code. Thus far, he has acquired a Hallicrafters HT-37, SX-111 with a matching speaker, Drake L-4B, and L7 linear amplifiers, MN-2000 antenna tuner, and TR4 Transceiver. At present, his latest project on the bench is a Hallicrafters SX-96 receiver. Restoring old radios is not any different from what gear-heads are doing with vintage cars. One of the commonalities between old cars and old radios is, as Bill already said, "you can still work on them!" If you want to take a ride down memory lane, find the *QST* article, *Washington Island's Ham Radio Treasure*, by Fred Lloyd, AA7BQ. It is in the Vintage Radio column of K2TQN on page 98 of the December 2011 issue. The subject is George Ulm, W9EVT (SK), and his collection located on Washington Island in Wisconsin. I had the pleasure of having dinner and a conversation with George at a W9DXCC banquet several years ago. What an interesting fellow. Check out Georges QRZ.com page at https://www.qrz.com/db/w9evt. As with the article about George, there is enough interest going on with vintage gear that QST continues to carve out space for a column every month. With a little digging, you will find regularly scheduled nets discussing vintage gear. Check out the ARRL net directory or the AC6V.com website and do a net search at http://www.ac6v.com/nets.htm#BA. You will find nets covering Collins, Drake, Hallicrafters, Swan, Heathkit, and others. Another interesting and useful link relating to all rigs is http://rigreference.com/en.

What do you do if you have a boat anchor and it has not been turned on for years, more likely many years? My suggestion is to wait for a moment. Unless you bring the line voltage up slowly, there is a good chance the electrolytic capacitors will fail to cause you an unnecessary mess to clean up. (***WARNING! You are dealing with LETHAL VOLTAGES! Do NOT attempt to work on the radio unless you have experience in doing so.***) If you are comfortable in opening the radio up, make a visual inspection of its interior, especially around the power supply area, and look for any leaky electrolytic capacitors. Assuming they look okay, what you want to do next is to find yourself an autotransformer, known well under the brand name *Variac*. This is where it is nice to belong to a local club. I am certain you can find a member who owns one and is willing to give you a hand in learning how to use it. Having secured one, determine the power rating of your radio while receiving. If it is a standalone transmitter, determine the power consumption while it is in idle mode, and ensure the autotransformer can safely handle it. ***(WARNING! Fire hazard—Do not leave your test area unattended while doing the following***.) Part of the testing process is to monitor the AC current being drawn while slowly increasing the voltage applied to the radio under-test. Insert an appliance ammeter between the 120VAC outlet and the device being monitored, in this case, the autotransformer. An alternative is to place a multimeter (set up for AC current), in series between the AC outlet and autotransformer. Now, before plugging the rig into the Variac, set it to zero, and then plug the radio into it and turn it on. Proceed by gradually turning the autotransformer up to 25VAC and checking to see how much current is being drawn. Is it high? Is it close to what should be drawn when operational or higher? If it is high, shut the test down. You likely have a shorted capacitor. If the current is low, let the radio cook for seven hours. (Depending on the amount of dust in the radio, you might experience a "hot" smell from the tubes heating up.) Now turn it up to 50VAC and again, check the current and if reasonably low, let it cook for another seven hours. As you have probably guessed, you are going to do this at 75VAC, then 100VAC, and then finally plugging the radio into a 120VAC outlet. Congratulations! I hope you did not experience any bad smell, smoke, or arcing, and you are hearing noise coming from the speaker. Although it is up and running, you still want to consider replacing those capacitors in the power supply, now knowing it is worth the time and money.

While you were poking around the club looking for that autotransformer, you may have come across a few of those chaps which I described earlier as being more interested in restoring old boat anchors than operating. I have a friend who fits that mold. As he, many work on only their gear simply for the fun of it. But then some are happy to help a fellow ham by doing a complete refurbishment of his radio; testing and replacing defective tubes, doing a complete alignment, and a good cleaning. I have met perfectionists who have gone as far as to search out original paint to redo cabinets. Some of them might only want to be reimbursed for the parts they replace, while some charge for their expertise. If you were to decide to pay for a refurbishment, I would lean towards finding a ham-technician who exclusively services the brand radio you have. I know for sure, some specialists only work on Collins or Drake, but it varies as far as other brands go. They are out there but you'll need to find them. This is where Google becomes your friend.

Well, so much for the boat anchors and memories past. What are hams using today? What happened after the dust settled in 1984?[73] As expected, solid-state HF transceivers and SSB became the new norm; all-tube and hybrid architecture faded away. The new transceivers no longer have tunable pi-output networks but rather a *no-tune* 50-ohm output. It wasn't long until 500-watt, solid-state, *auto-tune*, HF amplifiers were introduced by Yaesu, Kenwood, and Ten-Tec. Since then, 1500-watt solid-state amplifiers have appeared in the marketplace. HF Mobile transceivers became compact, and

Ham Radio is Alive and Well

Some of the latest technology with a little of the old. (W9RN, N. Amidzich)

Solid state XCVRS and amplifier. (W9MXQ, W. Shadid photo)

Dual receiver XCVR without a panadapter. (W9MXQ, W. Shadid photo)

remote antenna tuners started to arise. Direct Digital Synthesis (DDS) became available in ham gear, which gave us improved receiver dynamic performance and transmitter spectral purity. More importantly, it allowed synthesis lock when operating split and provided a quicker CW and QSK action (CW break-in). At the same rate, continuous frequency coverage emerged due to up-converting architecture. Because the WARC bands arrived during this period, all manufacturers started offering new HF models, with many offering upgrades to existing products. As DDS was a spinoff from the military, so became Digital Signal Processing (DSP). Audio DSP filters, heuristic noise reduction, and suppression of unwanted tones all became part of a new operating scheme for us. I revealed the impact the PC has had on the way we operate in earlier chapters. That came about somewhere between 1985 and 1995 and is when additional DSP techniques started to show up in transceivers. It was the Collins 75A-4 of 1955, and the Drake 1A in 1957, which originally brought us bandpass and notch filtering along with IF shifting. Some now refer to it as "Passband Tuning," but when the new transceivers had integrated IF-level DSP, those features truly stood out.

An operator owning one of these new radios can now easily adjust the IF bandwidth and shape factors at the turn of a knob. We should throw some digital noise reduction in there too. Moreover, all these functions are done within the AGC loop. Many of the radios available today may have a spectrum scope (panadapter) incorporated or at least an output to accommodate one. No different than our personal computers and smartphones, software has become an important part of our new radio. It is common to find regular software updates for a radio on a manufacturer's website. Because of the fixed 50-ohm output of the transmitter, many manufacturers added an automatic antenna tuner to their radio or made it available as an option. At the press of a tune button, the internal tuner will do a scan of your antenna system and match it to your transmitter. Although it is an antenna tuner, your expectations should not go beyond handling a mismatch of any more than 3:1. Internal tuners can adequately handle countless antenna systems, but not necessarily all. Any amateur using balanced-line, or who wants to match an exotic antenna, needs to purchase an external tuner. The other thing with later-model radios is that many do NOT have a built-in power supply. Having gone completely solid-state has allowed manufacturers to build radios which operate on 12VDC. A large number of 100-watt HF transceivers can essentially operate as mobile stations (certainly, there are variants among the brands). I personally prefer this configuration for a few reasons. First, the radio is lighter and easier to maneuver. Second, it makes it less costly to ship if the transceiver needs to be sent in for service. And lastly, the bulk of problems that occur in electronics is usually in the power supply and not the item being powered. Using a separate supply, not only allows us to own an overrated one for everyday use, but an extra one in case it fails. There are many good power supplies available in the market. Any capable of handling 25-amps or more should be fine for a 100-watt transceiver. The real upside to all this is that radio manufacturers are putting their time and resources into designing better radios and not wasting them on power supplies.

I am not going to attempt to explain the differences between the various brand radios, their functions, features, and benefits. You need to do that on your own. After all, that's part of the fun! You will find many first-class radios available from a variety of manufacturers. What we have going for ourselves today are the websites that render product reviews. One of the sites I enjoy using is http://www.eham.net/reviews/ (while there, you might want to check out their home page as well). The ARRL publishes product reviews in *QST* every month which are searchable on their website by members. The only advice I am going to give to those who are buying a radio is, "Do not buy your first one as being the last one." Many buyers go to spreadsheets and do comparisons of the brands and models trying to find that perfect rig that will last forever. Don't do it! I would first get involved with a club and see what other members are using. Look at QRZ.com pages and see what others have in their shacks. Go to some hamfests and conventions where equipment is on display. What rig looks fun to operate? By all means, looks are part of it. It is no different than buying a car. Do you see yourself sitting in front of it and having fun operating? My suggestion is to purchase a name brand, at the bottom or middle of the road in price, and get back on the air. If you are newly licensed, do the same and get on the air. Learn how to use that radio and make mental notes of what you like, what you wish it had, or what you hate about it. This is the radio, which you are going to use to set a baseline, or reference, to step up to your next or final radio. Personally, I had purchased three different radios before landing on that perfect one. Each one was a little bit bigger than the prior one and provided me with a few more benefits. I bought each one used except for the final one. The thing is, I was never dissatisfied with any of the earlier purchases. I continued upgrading only because I wanted to do more. In addition, it was not as if I was stuck with them. Between eBay and other ham radio websites, it is simple to buy and sell equipment today. My final purchase has been my best purchase and I am still satisfied with it. I knew what I was looking for because of what I used in the past; my reference. I do look at new technology when it arrives, but thus far have not found any new hardware that has tripped my trigger. The radio you choose continues to be dependent on what you enjoy doing in the hobby and obviously the size of your wallet.

The three big names and new mainstays in HF transceivers are Icom, Kenwood, and Yaesu. They all make fine equipment ranging from lower price products, $550 to $700, to the higher end of $7000 to $12,500, and everything in between. Go to *QST* and *CQ Magazine* and you will find a few other brands, especially in the VHF-UHF FM arena. Additionally, do not forget to check out eBay, Craigslist, QRZ.com, and QTH.com for used equipment. Many good deals turn up when other operators are upgrading to bigger and better. I purchased those first three radios of mine, a Yaesu FT-450AT, FT-950, and FT-2000, used and never experienced any problems. The same goes for accessories: speakers, headphones, power supplies, antenna tuners, rotors, and keys.

If you are returning from the "buy USA," boat anchor generation, as I had, you will appreciate this. You have heard the saying, "What goes around comes around," right? Well, unbelievably, the United States is making a huge comeback in the amateur radio market. There are two American brands, *Elecraft* and *FlexRadio Systems*, which are both leading the pack in new technology. As expected, it is neck-and-neck when it comes to the *"specsmanship."* It reminds me of the old audiophile days when there was a proven difference, but no one could hear it. Along with them are two other well-known American companies: Alpha RF Systems and TEN-TEC; both have had their share of bumps in the road, nevertheless they come across as doing reasonably well.

Software Defined Radio (SDR) is the latest in technology, and FlexRadio Systems happens to be one of the manufacturers who fall into that category. Recently, other SDR brands have been appearing on the horizon. They are Apache Labs, Odyssey TRX, and Expert Electronics, all made outside of the United States. Literally, the majority of these radios have no controls on the front panel except for an on-off switch, plus jacks for the microphone, key, and headphones. The rear panel looks akin to any other radio, having a full complement of jacks, ports, power, and antenna connectors. The interface to the radio is a PC monitor, keyboard, and mouse. If you are comfortable with and enjoy using, slider controls and buttons on a screen to control devices, an SDR is for you. Recently, in an attempt to lure knob turners such as myself, FlexRadio Systems developed an interface named Maestro®, whose face panel resembles a typical transceiver. The difference is that it is not physically attached to the radio. It is connected to the radio chassis-box via Wi-Fi or a network cable connection. The radio can be in one room while the control panel can be anywhere Internet access is available. More recently yet, FlexRadio Systems created a new package using the Maestro® front panel and a cabinet enclosing the

Gary L. Drasch

radio and connecting cables. This particular model certainty resembles a typical transceiver, as does the Expert brand. All the SDRs make operating remotely quite effortless.

It is human nature for us to want the latest and greatest in technology. Just because a radio might offer it, is not necessarily the best reason to purchase it. What normally comes with that greatness is a bigger price tag! There is a variety of incredible products available, but the price can get up there once they are fully configured. Again, decide what you want to do with your radio. If DX-ing or contesting sucked you in while using your first radio, you will likely head in the direction of the latest and greatest. However, if you were drawn into rag chewing and you are happy making QSOs regionally, do not waste your money. You might want to stick with that first or second radio you bought. The same goes for your antenna system. Because this is a hobby, you are allowed to spend as little or as much as you want. The main thing is that you have fun and ENJOY it!

Chapter 10
Antenna, antenna, antenna

There is nothing better than the smell of fresh aluminum in the morning. Undeniably, my XYL disagrees. You will find many more manufacturers of all sorts of antennas compared to the earlier years. The old HF mainstays such as Mosley, Hy-Gain, and Cushcraft have endured, but today they have competition from others, such as M2, Force 12, GAP, SteppIR, OptiBeam, Cubex, Acom, ECO, and JK. Along with those, VHF/UHF mobile and base station antennas far outnumber the HF antenna manufacturers. There are bunches out there.

The market which surprises me the most though is preassembled or manufactured wire antennas. I remain in disbelief. Really? Preassembled wire antennas? The answer is yes. Companies such as Alpha Delta, Array Solutions, Buckmaster, B&W, Buxcomm, MFJ, and Radio Works, manufacture dipoles, Carolina-Windoms, fan dipoles, loop, and end-fed antennas. All are cut to length with a balun (balanced-to-unbalanced device), coax connector, and insulators all fittingly attached. To my amazement, they are being sold like crazy. I've even heard users in QSOs talk about the brand name dipole or end-fed antenna they are using as if there is actually a difference. Better yet, these wire antennas are written up in product reviews. Come on—they are joking—right? Are hams no longer capable of buying wire, and insulators, and building their own anymore?

I love building and experimenting with antennas, from wire to some of the simpler beams and verticals. Antennas are another one of those disciplines of a hobby within a hobby. The resources available these days compared to the 1960-70s are unbelievable.

Gary L. Drasch

My KT-34A Tri-bander at 38' and 6m homebrew at 46'. The tower is also gama-matched on 160-meters. (G. Drasch photo)

There is so much to learn and experiment with when it comes to antennas. Pick up a copy of the *ARRL Antenna Book* if you don't believe me. By doing so, you will be pleasantly surprised to find you also received access to the fully searchable digital edition of the printed book, as well as utility programs and supplemental content including expanded technical papers, in-depth construction guides, and referenced articles.

As far as building materials go, there are plenty of sources. Just Google "antenna aluminum tubing for ham radio" or "antenna wire for ham radio."

You will uncover all the materials a ham needs to build whatever type of antenna he or she wants. The same goes for various clamps, stainless steel hardware, cabling, connectors, towers, and masts.

Don't forget to peruse the assortment of online utility programs that comes with the *ARRL Antenna Book*. One of them should be a copy of *EZNEC*. It is a tool I simply find mind-blowing. EZNEC is an antenna modeling program that permits users to model an antenna, from a dipole, loop, vertical, or beam right on their computer. They enter the data such as element (tube or wire) length, diameter, spacing, height above ground, loading coils and traps, and even the type of soil. Any parameter, which affects the performance of an antenna, can be entered. The program is not very intuitive and therefore it makes sense to pick up an ARRL copy of *Antenna Modeling for Beginners* by Ward Silver, N0AX. Ward systematically takes the reader into the modeling of their first antenna. It might come across as a little laborious to the user until they become comfortable with it. Once they do, it will provide them with both azimuth and elevation radiation patterns, DB gain, the complex impedance, and the Standing Wave Ratio (SWR). They may ask "what if" questions. How does it perform over Real Ground compared to Free Space? Is it worth raising the antenna ten or twenty feet? Does running the wire at an angle make a difference?

The EZNEC program in fact works. For example; I acquired nearly all of the pieces of a four-element 6-meter beam from a friend a while back. I took it home, cleaned it up, and reassembled it to what I thought was correct. I was not exactly sure what brand it was, but it looked similar to a Hy-gain VB-64DX. I found the manual online and downloaded it. To work on the antenna, I placed it onto some homebrew holders fastened to a pair of old sawbucks. The anticipation of having a 6-meter beam was starting to fire me up. I connected my MFJ-259B antenna analyzer to the relic and found my excitement quickly dashed. The impedance was high and I had a 3:1 SWR. I kept dinking around with the coaxial matching loop thinking it had to be the problem. That wasn't it. It was a head-scratcher.

Here I had built it as close to the specs in the Hy-gain manual as I could and the darn thing wasn't working. As so often, I brought the subject up to Lyle over pizza one night. He asked me, "Did you model it?" My response was, "no," of course. The next morning, I began entering all the data, i.e., the real measurements taken directly off the beam—not from the manual. When done, I ran EZNEC, and lo and behold the results matched my antenna analyzer to a tee. They were identical, same impedance and SWR. I was elated! This exercise did not only give me faith in my analyzer but also in the software. So, how was the construction different from the drawing and measurements I was using? During the measurement process, I did happen to notice the diameter of the driven element was at least twice, if not more, the diameter specified in the Hy-gain manual. The element-to-boom connection was a Cushcraft design using those large black insulating sleeves, slid over the driven element, and then clamped between special element-to-boom connectors. I thought, "Can that be it, the element diameter?" It surely was. The EZNEC program proved it. I was grinning ear-to-ear and immediately started remodeling the antenna using the same size diameter elements but instead; I adjusted their length within the model. There it was; an antenna model with new element dimensions which should work. The program model showed a 50-ohm impedance with a 1.1 SWR across the portion of the 6-meter band I was looking for. It was time to replicate the program measurements to the actual beam. I needed to find some tubing to fit inside the existing elements to lengthen each one a little bit. After making the physical adjustments, my antenna analyzer now read 50-ohms with a 1.1 SWR. It was another case where the PC and program enhanced the hobby.

I recently added an additional piece of equipment to my arsenal. Its called a NanoVNA (Nano Vector Network Analyzer). In the past, I used to plot my SWR vs frequency on graph paper. It entailed adjusting the frequency of the MFJ-259B for each point I wanted to highlight on the graph—doable but a pain in the butt.

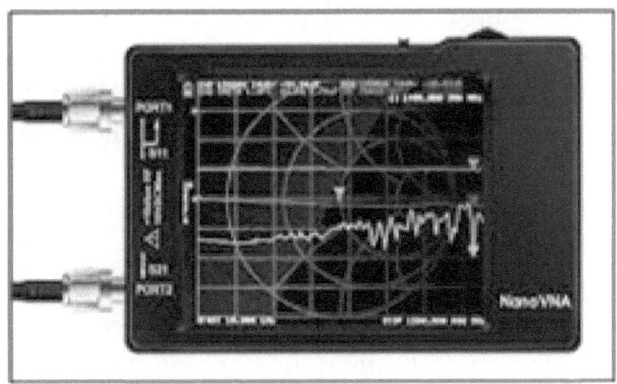

NanoVNA- 1.5GHz (Amazon photo)

The correct way has always been to use a VNA, but the cost of one in the mid to late 1980s was in the $100,000 category. Thanks to new technology, and of course the use of a PC, it's now possible to buy one in the $45 to $140 arena. Unbelievable! A person needs to go online to download the *"NanoSaver"* software at https://nanovna.com/?page_id=90. Yes, it's free. With all things being relative, I had found it quite intuitive. Using the NanoVNA as a standalone device is another thing though. For my part, it wouldn't need a display. I constantly use mine connected to the PC and software all the time. The VNA is so much more than just an antenna analyzer—so much that it is beyond the scope of this book. Learn how to use it, and It will actually become an excellent learning tool.

The ARRL did a great job creating the *Antenna Book*. The EZNEC program and ARRL YW (Yagis for Windows) software can make use of various antenna models straight out of the chapters of the book. That is the main reason the ARRL makes it available for download with purchase. You may learn more, and order EZNEC, at http://eznec.com/. It is now free!

Do you remember the old saying, *antenna, antenna, antenna*? Nothing has changed much in that respect. One's antenna system continues to be the crucial building block of the performance of any station, may it be transmit or receive. We can own a small affordable low-power QRP station connected to a high-performance antenna system and do quite well. As an example, I was testing an antenna switching system I had built, and found an unused frequency on 20-meters and turned my power down to ten watts. I only wanted to take a quick look at my SWR without necessarily making a contact. I identified myself and said, "test, test, this is K9DJT," and let go of the PTT (push-to-talk) floor switch. So, what happened besides my SWR being fine? Some bucko comes back to me and says, "You're five-by-nine in Florida." I'm thinking, this is unbelievable, I am only running 10 watts. I had not given any thought of being connected to a four-element tri-bander pointed in his direction. Clearly, band conditions played into it too, but nevertheless, 10 watts SSB? Not only was I using a beam at 38 feet, but I was also using a 7/8" Heliax transmission line to feed it.

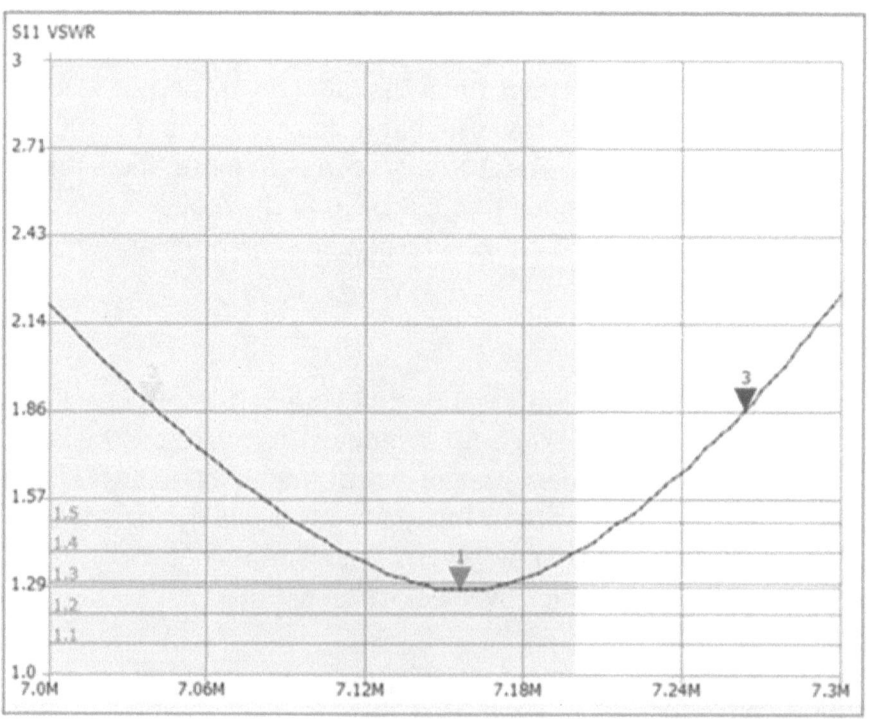

VSWR plot of my 40m half-sloper antenna using "NanoSaver" software. (G. Drasch photo)

That particular coax, at 14.2-MHz, has less than .16-dB loss per 100-feet, which means it's almost as if my radio is right at the antenna. I experienced a similar QSO with a gentleman in Texas early one evening on 6-meters. He said my signal was booming. I was running 100 watts into my 4-element beam at 46 feet pointed at him.

20-meter, 3 stack,6 element, OWA Yagi (K9CT, C. Thompson photo)

(I was using the same 7/8" Heliax cable as on 20-meters.) He and I decided to conduct a little experiment with me gradually decreasing my power. Long story short, I descended to 1 watt and I retained my 59 signal with him. It was really a very memorable QSO experience.

The moral of the story is to create an antenna system as efficient as possible. That means a resonant antenna along with a low-loss transmission line. An antenna is only resonant at one frequency, so we want to tune it to somewhere in the middle of our frequency range of interest. If our SWR is higher towards the ends of our selected range, or we are using a balanced transmission line, we want to use an antenna tuner to match it to our transmitter.

Speaking about antenna tuners—somewhere along the line you may have heard an operator say, "I have an antenna tuner that will tune a wet noodle." That is marvelous, but it does not mean he has an effective system. All the tuner is doing, besides keeping your transmitter happy, is ensuring all the power is going to a wet noodle; it does not mean it is being radiated. Many refer to this as a conjugate match. It is when a tuner matches the complex impedance of whatever you are attempting to use as an antenna to 50 ohms. The whatever might be a random length of wire, a light bulb, or even a dummy load. They will all work, and all of them will radiate energy. But the thing is, they are not very capable antennas. That low SWR between your antenna tuner and the transmitter doesn't mean you have a perfect antenna system. A low-loss transmission line and a resonant antenna bring it closer to reality. Understanding a resonant antenna is one thing compared to understanding transmission lines and SWR. One of the best articles which helped me understand SWR was written by Darrin Walraven, K5DVW, titled *"Understanding SWR by Example,"* published in the November 2006, *QST*. It is a must-read.

An efficient antenna does not necessarily need to be expensive. My original antenna was a dipole, which by the way, I designed and built myself. I strung it between a utility pole in my backyard and the light pole across the street and supported it in the middle with some PVC pipe. It worked fine. I have several friends who have similar antennas and have not gone any larger. The antenna is fulfilling their needs. They primarily check into regional nets and rag chew around the country, even making occasional DX contacts here and there. In a couple of cases, they are limited in what they can put up in their neighborhood. Some radio amateurs have gone as far as to create a stealth antenna, using small gauge wire in order to camouflage it. Others have kicked it up a notch. Look around your neighborhood—that flagpole down the street might actually be a vertical antenna. Antenna interests change depending on what hooked us in the hobby. I evolved from being happy making a few QSOs to obsessively chasing DX. So obsessively, the simple dipole I had been using mushroomed into a mini antenna farm on a city lot. Even with that, it doesn't need to be ultra-expensive. For example, I have three towers on my property, with two rotors, a tri-bander, a 12/17-meter beam, a 70-cm, a 2-meter, a 6-meter beam, and various wires. My wire antennas are all supported between the towers and the same light pole across the street.

Would a Rohn 25G tower be of interest to you? There is a supply out there just for the asking and removal of it. People are no longer using the old residential TV towers because of the advent of cable and satellite TV. I acquired all my towers by offering to take them

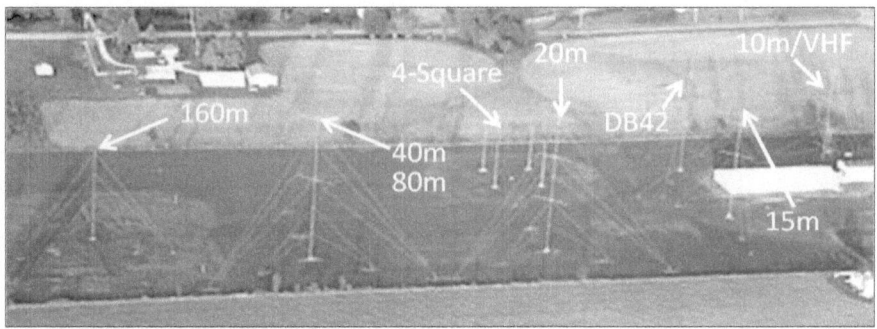

K9CT Antenna Farm - And Yes, those are individual towers. (WO9W, W. Cusack photo)

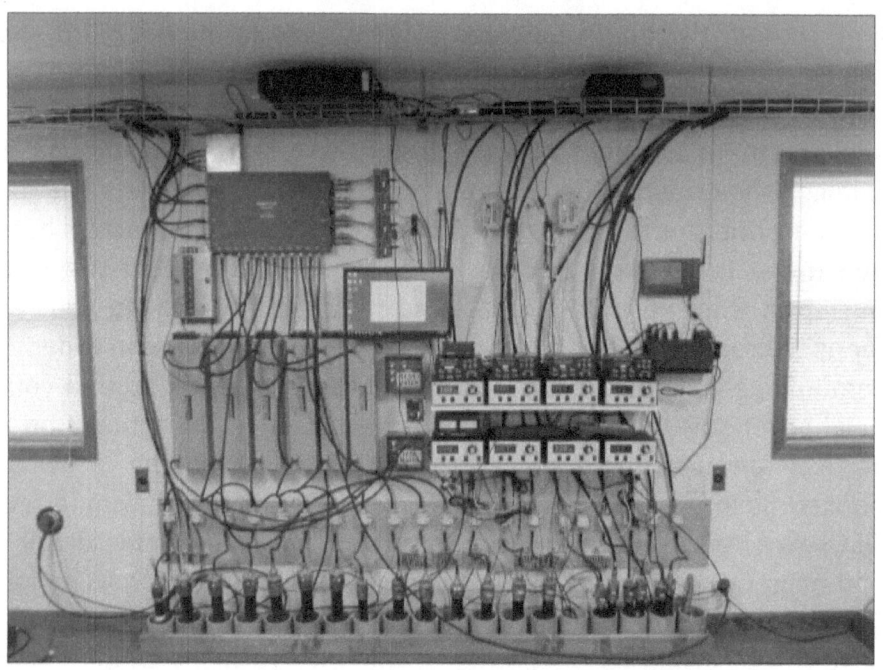

K9CT antenna switching and rotor controls.(K9CT, C. Thompson photo)

down for families. I even have a fourth which is ready to take to my cabin. Overall, people are pleased as punch to have them taken down and be gone, without paying for removal. My tri-bander and both rotors, I purchased used. Both the 2-meter and 70-centimeter beams were given to me by the wife of an SK, again, for taking them down. I did buy new aluminum for my homebrew 12/17-meter beam though.

Okay, the above is all fine and good for an old codger such as I, who likes to climb and is willing to disassemble and reinstall towers and antennas. What about the OM who dislikes heights nor has the mechanical skills required to take down a free tower? He should hit the hamfests. It is common to find used, inexpensive towers at the ones held during the summer. As far as the climbing goes, I trust there is a young do-all within a local club, or at least in a club not too far away, who would be willing to earn a little extra cash. As a last resort, a professional climber/installer can be contracted. My friend Lyle continues to climb, but when it came to putting up some used self-supporting cell phone towers, he contracted a professional.

I have paid for used Heliax cable and connectors but that is where it stops. It is another story when it comes to any other type of transmission line and/or connector. We want to buy the best quality available. A well-known contester in one of my clubs once told me, "A cheap connector is an easy way to wreck a weekend." Of course, I did not heed his advice and needed to prove it to myself. All connectors might look the same in those eBay ads, but believe me, they are not. The same goes for coax. Always buy a brand-name cable and connector. It is worth the extra money in the long run.

Were you curious earlier when I spoke of a balun on a dipole? The use of a balun was not prevalent during the 1960s and earlier. We are now using these on the coax of not only dipoles but all of our antennas fed with coax. Over time, the communications industry discovered coax does not have currents flowing only in two directions such as a balanced-line, but rather three—one on the center conductor, another on the inside of the shield, and an additional current flowing on the outside of the shield. Instead of connecting our coax to the center of the dipole and letting it hang, we now add several loops of coax right below the connection. Some of the beam manufacturers started doing this in the mid-1960s to early 1970s. The more popular alternative today is to add a bead-balun at the feed point of our antenna, i.e., several ferrite beads slid over the coax at the antenna connection. The main concept of adding a coiled coax or ferrite beads is to eliminate that third current flowing on the outside shield of the transmission line. Imagine the havoc a radiating shield can cause with consumer electronics to problems in our own shack. To learn more and to get the real technical rundown, reference either an *ARRL Handbook* or *Antenna Book*. It is well worth the study.

Gary L. Drasch

I have a few friends whose opinions are a little skewed as to when and how well to do antenna work. Reasonable hams agree with the idea of working on their antenna system during decent weather or at least before the snow flies. That in effect is the plan for us living in the upper Midwest. The thing is—I still manage to make repairs during the winter. I once climbed my tower in February a few years ago to replace a sheared-off bolt on my rotor (the beam was turning freely all on its own). A couple of years ago I replaced a connector on a transmission line in 10-degree Fahrenheit temperatures because water made its way through the electrical tape; it froze and broke the PL-259 connector apart. Some of my buddies say, "Winter is the best time to do antenna work!" However, if you ever met any of my buddies, you would understand. Winter antenna work does not particularly thrill me. The cable is stiff, the tools are ice cold, and nothing is ever large enough to work on while wearing gloves. The best is when I drop a tool or a connector in the snow and cannot find it. Then to top it off, when I go inside to warm up, I discover little cracks, almost like tiny cuts, on my fingers from the cold. They are usually near a fingernail or on the knuckle, and always hurt like hell.

So why do we punish ourselves? Repairing the rotor was to enable me to work a rare DX-pedition. And I did! The coax connector was to work a contest, which never did come to fruition because I had additional outside problems. I think the real reason is that we need to suffer to make up for our own stupidity. After all, most of our problems are self-inflicted. If I had used a hardened-steel bolt instead of a soft one, it would not have broke. If I had used quality electrical tape such as Scotch 33+, rather than some cheap stuff, water would not have gotten into the connector. Better yet, if I had replaced the cable rather than spliced it, there would not have been any downtime or problem at all. I get down on myself when things break down and realize it could have been prevented. I'm always trying to improve my workmanship, or at least anticipate what can happen, so I don't need to address antenna issues in the winter. One morning during a QSO with a friend of mine, Ron, W9BCK (SK),

I began explaining my professionalism, which I was implementing during the installation of a new antenna. To my surprise, he replied, "no, no, no. You are wasting your time." I did not understand. "What do you mean?" I asked. Ron said, "At some point, you're just going to change things again. Why bother?" You need to know Ron. He is an experienced man who is highly regarded not only by me but also by many in our local club. Being in his late eighties at the time and saying what he did broke me up. I could not stop laughing. Here I was, expecting to impress this retired broadcast engineer with my quality workmanship, and he told me I was wasting my time. I replied, "Really?" He came back and said, "Yes, you need to operate as though it is a permanent Field Day. You are always going to be changing something." Although I persist in doing things right, something my grandfather taught me, Ron is as good as on the money. I am unable to count how many times I have changed things around with my various installations.

I take tower safety very seriously. I make use of a fall protection harness and follow all the necessary safety precautions except for one. I have a habit of climbing my towers while my wife is gone. She gets upset with me saying, "what if something happened?" I guess I always count on there being a nosey person in the neighborhood who might be watching me and would call 911. I shared my wife's concern with my friend Tom, W9IPR, at a club meeting one night. Tom replied, "There's nothing wrong with that. I always climb my tower while Pat is gone. That way if I fall, I don't need to listen to her bitch at me!"

You know, I think he might have something there!

Gary L. Drasch

The antenna farm of OM8A, Slovak Republic (W0GXA, R.C. Lee photo)

Chapter 11

Beyond Lids

This is a sore spot for myself and the entire ham radio community. It is the reason I placed this chapter towards the rear of the book. Fortunately, only a few people fall into this subject matter and spoil it for the majority. If you are unfamiliar with the term Lid, it means "poor operator."

A DX-er refers to an operator who parks on a DX transmit frequency and reprimands others as a cop. This is the frequency where the DX station responds to callers who are transmitting on a different frequency up from his. This self-appointed cop sits there, just like a squad car with radar, waiting for "Mr. Non-split" to screw up and call on the DX's frequency—the wrong frequency. He cannot wait to shout at the offender to let him know. Preferably, there would be no cops, but there are a few individuals who look upon themselves as doing the DX-ers a service. As I said, these cops are self-appointed—they are not members of the DX-pedition, and they become nothing more than an annoyance. They naturally show up during all the large and rare DX-peditions, and there is always enough of them to cover all the bands. No one understands what pleasure they get out of it other than feeling superior by knowing where the DX is listening. Unfortunately, they are here to stay. The fact is, every time a cop corrects a wrongdoer, he himself is creating interference for the DX-ers by covering up the very station the non-offenders are trying to copy. There is not a DX-er alive who has not called on the DX's frequency by mistake. A true DX-er knows the DX is listening up, but for an unknown reason forgot to place the radio into split mode. I did it once. I was attempting to work a DX station with my dual receiver radio, listening to the DX in my left ear and the pileup in my right. I knew where I wanted to be transmitting, and in fact, thought I was on the correct frequency—the one above the DX—the one in my right ear. As luck will have it, I forgot to split my transmit VFOs. The result was a reprimand from a cop saying, "Up Stupid." Hey, I am not stupid; I made a mistake! On the other hand, he might have said, "up...up...up...up!" Think about it. First, my mistake covered up the DX station for two seconds, and then the cop covered the DX for another one to two seconds or more. Between my error and the cop, we may have busted an ATNO of a fellow DX-er. I will say this though; I made that mistake only once. I was truly embarrassed.

In all societies, there are good cops and bad cops. What is the difference between ham radio? A good cop only says, "Up!" It's simple; he makes the point and uses the least amount of time. That is the important part, the least amount of time. The corrected operator can be polite and say, "Thank you" to the cop, but that is not being a good operator. A good operator doesn't waste any more time than he and the cop already have in covering up the frequency. A good operator immediately splits his VFOs and gets on with working the DX. Another thing a good operator does is NOT respond to a cop's derogatory remark. It doesn't make any sense to reply to the officer who called me stupid with, "who you calling stupid scumbag?" Regrettably, I have heard that done, and then an argument ensues and covers the frequency even longer. If I did that, I would be no better than the cop who called me stupid in the first place. I merely left and disappeared into the pileup.

The thing that irritates me the most is deliberate interference, or what is called direct QRM (DQRM). I hear schlups intentionally transmit nonsense on the DX's frequency so others cannot hear him. For example, they will lay a carrier (an uninterrupted CW signal) on the frequency, or maybe send a series of dashes or dots for minutes at a time. They might send a digital signal for long periods. I even heard some fool playing music in the CW portion of the 40-meter band one evening. They do this to spoil the fun of others. They consider it funny. This sad behavior is not limited to

DX-ing. I have a friend who told me how people have been causing DQRM with nets as well. I am yet in disbelief and do not have any recollection of comparable behavior going on like this in the 1960s.

So, how do we handle the situation? I have wanted to give a few jerks a piece of my mind more than once when I became outraged by their conduct. Instead, I chose to push back from my operating position, took a deep breath, and turned the radio off. It was not worth letting them get to me. I have had several discussions with members of the DX club I belong to, and found it was unanimous; the best way to deal with a QRM-er (a person intentionally causing interference) is NOT to respond. What fuels the yo-yo is replying and letting him know he's bothering you. That is exactly what he is looking for. It makes him happy when he knows he's getting under our skin and will rudely continue. He is not having any fun when he discovers he's not causing problems. If we ignore him, he will sooner or later go away.

During the Centennial Points Challenge, I made contact with Kay Craigie, N3KN, who was the president of the ARRL at the time. Being president meant extra points for participants who worked her. Do I need to say more? As you would expect, someone was causing DQRM. Kay was barely readable but I was totally ignoring it. Kay made a comment about the interference and I replied saying, "It's not a problem, that's what notch filters are for" and unbelievably when I returned it to her, the guy was gone. That's the key. As annoying as it is, do not engage with the individual. Ignore him or switch to a different band. Change things up for yourself, or turn the radio off and come back later.

I discovered the day of the week, and the time of the day seems to be a factor in when DQRM occurs. Being retired allows me to operate whenever I want, which in turn permitted me to make this surprising observation. DQRM is a worldwide problem, but it appears the DQRM-ers in the United States are the working class, meaning the non-retired. The majority of the DQRM during the week is from late afternoon to midnight. Sad to say, we can also count on it occurring from late afternoon on Friday to midnight on Sunday. There is next to no DQRM during the weekday, being non-existent from midnight until late afternoon. It appears the few misguided people causing the DQRM are jobholders by day and QRM-ers by night and weekends.

Then there is Slim. One of my favorite and memorable DX-peditions was the Amsterdam DX-pedition, FT5ZM, in February of 2014. Band conditions were extraordinary and I was having a ball filling the band slots. I was working them on all the bands one right after the other. They happened to have an online log they updated regularly and I checked daily. One night when reviewing it, I noticed I was not in their log for my 40-meter CW contact. I'm thinking, how can it be? If I had ever made a solid contact with a DX-pedition, this was it—hands down—no question about it. Nevertheless, I wasn't in the log. I brought this up to Lyle, WE9R, over pizza at our local meeting spot and he started grinning. I said, "What?" He said, "You worked Slim." My response was, "who in the hell is Slim?" He explained that Slim was another one of those misguided souls who pretends to be the DX station. He gets on the exact same frequency, turns his power way down so he is as weak, and tries to match the same code speed of the DX station. I know, it is hard to believe.

So far, I have only managed to work Slim on CW. Does he attempt the same stunt on SSB? No, if he did, all of us would hear him laughing! Slim is a tough one to get around unless we can hear the difference in the signal strength between him and the real DX station. He will get on a roll once the pileup starts answering him, and when that happens, many operators will have wasted time chasing a ghost. The worst of it is no one knows until they do not find themselves in the logbook. This is why online logbooks, uploaded by the DX-pedition, are so helpful. It gives us a second shot at the DX-pedition as long as they haven't packed up. I expect Slim holds down a job similar to the QRM-ers because he isn't around during weekdays either—only at night. I do get upset when I find out I've been drawn into his game though. The only way I justify my time working these space cadets is by looking at it as practice in working a pileup. After all, he did call me!

I have heard fellow hams discuss bad behavior and say, "That's what happens when you drop the code requirement," referring to licensing. That's not it. All of the Slims are good CW operators who copy at a high rate of speed. They are not new to the hobby. Reinstating a code test will not fix the Slim problem. As far as the DQRM goes, would a code test fix that? I am not sure. What it comes down to is the behavior of some hams is no different than a select few in any society. We are not immune to those who partake in deplorable behavior. So what is the solution? Maybe we just need to wait. As my friend once said about his really bad neighbor, "She'll eventually die."

Gary L. Drasch

Chapter 12

Ham Radio Speak

I expect you are still familiar with much of the slang and abbreviations, even if you've been inactive and are now just returning to the hobby. You may want to skip this chapter, then again, it might be a good review. Maybe there are a few new terms; or you'll say to yourself, "Oh yeah, I remember that!" My main goal here is to provide the reader with an idea of how hams communicate not only on the air but also in general.

Along with slang expressions come abbreviations, prosigns, and brevity codes. The difference is, an abbreviation and prosign are shortened forms of a word or procedure, while a brevity code is used in reducing the amount of time it takes to say a sentence. As young people created their own form of language for texting, ham radio operators had done the same, but they did it much sooner, like multiple generations sooner.

The majority of abbreviations, in contrast to slang, are used in ham radio while sending and receiving the Morse code. Railroad telegraphers created a bucketload of abbreviations, which modern code operators continue to embrace and use today. The popular abbreviations are listed in the table below. Keep in mind the list is only a partial of what is in use.

Prosigns are referred to as procedural messages which use a special dit-dah sequence. They do not represent any specific word but rather a non-language function relating to the exchange of information while sending and receiving Morse code. The prosign list follows the abbreviations list below.

There are two groups of brevity codes used in ham radio. One is the Q code, started in 1909.[74] It is otherwise known, or referred to, as Q-signals. The Q-signal may either be a statement or a question. Sometimes there might be a slight variation in Q-signal usage, which I show in parenthesis in the table below. The second group is the RST code, widely used by 1912,[75] and solely referred to as RST. The reason these came about was to create a shorthand for the commercial Morse code operator, later continuing into CW radio communications. Because they were so well known and widely used by CW operators, it wasn't a surprise they were carried forward into phone communications. The RST code is a system, which uses numbers, to provide a comprehensive signal report to a station an operator is working. RST stands for Readability, Strength (signal), and Tone (or Quality). The Quality is used in place of Tone when operating digital modes such as PSK or RTTY. Only readability and signal strength are used when operating phone. Discussions have been taking place regarding the tone portion of the RST report; the question is, is it still of value? During the early years, when transmitters were primarily homebrewed, there was a significant difference in the tone quality of a CW signal. That signal was dependent on the quality of the equipment the operator built. Currently, the chances of coming across a homebrew transmitter and power supply combo are slight. Today, the majority of rigs are commercially manufactured and they all sound good. Clearly, there are boat anchors out there that press on and generate some questionable signals. Only time will tell.

Below is a list of common Slang[76] expressions used among radio amateurs on the air, face-to-face, emailing, and texting:

- 807 – Old glass vacuum tube several inches tall; Slang for a beer, as, "let's get a couple of 807's."
- Alligator – Operator with lots of power and no receiver, i.e., a big mouth and unable to hear.
- Antenna farm – Multiple, large antennas, owned by a ham.
- ATNO – All-time new one. A country never worked before.

- Band points – Unit of measure used with the DX Challenge Award.
- Barefoot – Running a transmitter without an amplifier.
- Beam – A multi-element antenna typically located on a tower.
- Big gun(s) – High-powered station(s).
- Bird – Amateur radio "satellite."
- Blue Whizzer – Rare ionized meteor trail lasting several seconds or longer.
- Boat-anchor – Old, vintage, or classic radio equipment.
- Brick – Any group of electronic components encapsulated into a block.
- Bug – Semi-automatic Morse code key using a spring action to create dits (dots).
- Bureau – A clearinghouse, and routing system, for QSL cards.
- Cans – Headphones.
- Chirp – CW signal with a bird-like tone.
- Cloud Warmer – An antenna that radiates straight up.
- CW – Continuous-wave signal; also meaning Morse code.
- D-Star™ – Digital Smart Technologies for Amateur Radio: Icom's digital communication protocol.
- DIT DIT – The sound of two E's in Morse code. Means and sounds like "bye-bye."
- DMRTM - Digital Mobile Radio: A commercial radio digital communication protocol.
- DQRM – Deliberate interference.
- DX – Distance Stations.
- DX-er – A ham looking for DX.
- DX-ing – The act of looking for DX stations.
- DX-pedition – Group of hams who travel to an exotic location to specifically activate for others.
- Es – Sporadic E propagation: metallic ion layers using the ionosphere.
- Elmer – Mentor and helper to new radio amateurs.
- EmComm – Emergency communications

- EME – Earth-Moon-Earth (moonbounce)
- Eyeball – A face-to-face conversation.
- Fist – The unique sending style of a particular Morse code operator.
- Flagged – Being heard on PSK Reporter or WSPR
- Gentlemen's Band – 160-meters (1.8-2.0 MHz).
- Grid-pedition – One or two hams who travel to a rare grid location to specifically activate for others.
- Harmonic – Children, e.g., I have two harmonics, a boy, and a girl.
- HF – High frequency (1.8 to 30 MHz).
- Hollow state – Tube equipment.
- Homebrew – Homemade equipment.
- Ionoscatter – Scattering of RF in the ionosphere D region.
- Ionosonde – Special radar to examine the ionosphere; Its signal is often seen sweeping across an HF bandscope
- Key, Straight Key – Morse code lever type of device.
- Keyer – Electronic circuit, which creates DITs and DAHs depending on which key paddle is depressed.
- Lid – Poor operator.
- Little pistol – Lower power station
- LoTW – Logbook of The World.
- Machine – A RF repeater.
- Magic Band – 6-meters (50-54 MHz).
- Matchbox – An antenna-tuning device between the transmitter and an antenna.
- MS – Meteor Scatter.
- OM, Old man – Any male ham operator.
- Opening – The atmosphere and/or solar activities, is supporting communications.
- OQRS – An Online QSL Request Service.
- Paddles – An Iambic type of key.
- Phone – Voice communications, using a microphone.
- Pileup – Many stations all calling a single station at the same time.

- Pirate – An operator using either someone else's call sign or a made-up one.
- PJ – PingJockey: the name of a bulletin board used while operating meteor scatter.
- Pounding brass – Operating using Morse code.
- Q – A QSO.
- Qs – The number of QSOs (contacts) made during a contest or period of operation.
- QRM – Interference
- QRM-er – Person causing intentional interference, DQRM.
- QSL-ing – The exchange of confirming QSLs (postcards).
- Rag chewing – Talking about anything and everything.
- Reading the mail – Listening to a QSO/conversation without participating, i.e., without transmitting.
- RF – Radio frequency.
- Rig – The radio amateur station; an operator's equipment.
- Rock – a quartz crystal used for the frequency control of a transmitter (primarily used years ago).
- Rock-bound – a transmitter, which can only use crystals (a rock).
- Rox – Meteors
- Rover – A portable station used during a Grid-pedition.
- Rubber Duck – Also known as a "Rubber-Ducky." A flexible antenna found on walkie-talkies.
- Running – 1.) Calling CQ in a contest on the same frequency for an extended amount of time. 2.) When both stations of a meteor-scatter QSO begin to decode each other.
- Shack – The room where the radio equipment resides.
- Shout – To call another amateur on the radio.
- Simplex – Receiving and transmitting on the same frequency.
- SK, Silent key – Refers to a fellow ham who has passed away. His key has gone silent.
- Skimmer – An automated CW/RTTY spotting device (cluster).

- Slim – A CW imposter pretending to be a DX station. Also known as a pirate.
- Slack – the name of a bulletin board used while operating VHF/UHF, Rovers, EME, and meteor scatter.
- S & P – Search and Pounce in contesting.
- Split – Transmitting on one frequency while listening on a different one.
- System Fusion™ – Yaesu's digital communication protocol.
- Tail-ending – Interjecting your call sign on top of the non-DX station just before he ends his transmission. (Not cool)
- Tail-gating – Transmitting your call sign just as the pileup dies down, but still while the DX is listening.
- Ticket – Radio amateur license.
- TNC – Terminal Node Controller
- Top band – 160-meters (1.8-2.0 MHz).
- Traffic – Information passed between various stations.
- Tri-bander – A three-band beam antenna (10, 15, and 20-meters).
- Tropo – Tropospheric scatter: VHF & higher frequencies pass through the upper layers of the troposphere.
- Wallpaper – Awards and certificates.
- Work or worked – Make or made contact with. Example: "I worked K9VNM on 17-meters."
- XYL – Wife (ex-young lady).
- Yagi – A directional antenna consisting of two or more elements.
- YL – Young lady; any female ham operator.
- Zed – A way of pronouncing the letter "Z" to eliminate confusion with the letter "E."
- Zero-beating – Matching the transmitter frequency to the receiver frequency.

This is a partial list of the commonly used CW abbreviations:[77]

- 73 – Best Regards.

- 88 – Hugs and Kisses.
- ABT – About.
- AGN – Again.
- ANT – Antenna.
- ARND – Around.
- B4 – Before.
- BK – Break, Back.
- BN – Been.
- BTR – Better.
- BUX – Dollars.
- CLDY – Cloudy.
- CLR – Clear (usually a description of the weather).
- CNTCT – Contact.
- CONDX – Conditions.
- CPY, CPI – Copy.
- CQ – Calling any station that can hear me, I am looking for a conversation with anyone.
- CU – See you.
- CUD – Could.
- CUL – See You Later.
- CUZ – Because.
- DE – From (or this is).
- DR – Dear (often used by foreign/DX operators in front of the other operator's name).
- ES – And.
- FB – Fine Business (OK, good, terrific).
- FER – For.
- FT – Feet.
- GA – Good Afternoon.
- GB – Good Bye.
- GE – Good Evening.
- GL – Good Luck.
- GM – Good Morning.
- GUD – Good.
- HI HI – This is the way a laugh is sent via CW (Morse code).

- HPE – Hope.
- HR – Here, Hear.
- HV – Have.
- HW – How.
- NR – Number.
- NW – Now.
- OM – Old Man.
- OP – Operator.
- PSE – Please.
- PWR – Power.
- R – Are or Received.
- RCVR – Receiver.
- RPT – Report.
- RST (Q) – Code for Readability, Strength, and Tone (Quality).
- SIGS – Signals.
- SN – Soon.
- SRI – Sorry.
- TMW – Tomorrow.
- TNX, TKS – Thanks.
- TU – Thank you
- TX – Transmit.
- U – You.
- UR – Your, You Are.
- URS – Yours.
- VY – Very.
- WTS – Watts.
- WUD – Would.
- WX – Weather.
- XCVR – Transceiver.
- XMTR – Transmitter.
- XYL – Wife.
- YL – Young Lady.
- YRS – Years.

The following prosigns,[78] are used when communicating in Morse code. They use a special DIT-DAH sequence which are considered procedural messages:

- AR – Over, end of message.
- K – Go, any station is invited to transmit.
- KN – Go now, only a specific station to transmit.
- BK – Invitation for the receiving station to transmit.
- R – Received fine.
- AS – Stand by, please.
- SK – End of contact.
- EE – Bye-bye.

You should find this list as being the common Q-Signals used on the air:

- QRG – Will you tell me my exact frequency? Or, your exact frequency is _____ kHz.
- QRL – Are you busy? Or, I am busy. (Is this frequency busy or in use?).
- QRM – Is my transmission being interfered with? Or, your transmission is being interfered with.
- QRN – Are you bothered by static? Or, I am bothered by static.
- QRO – Shall I increase power? Or, increase your power.
- Also denotes the use of a high-power amplifier.
- QRP – Shall I decrease power? Or, Decrease your power. Also denotes the use of a low-power transmitter.
- QRQ – Shall I send faster? Or, please send faster.
- QRS – Shall I send more slowly? Or, please send more slowly.
- QRT – Shall I stop sending? Or, please stop sending. Also means I am shutting down my station.

- QRV – Are you ready? Or, I am ready. (Also indicates a new station is ready to operate, e.g., the first transmission after setting up a DX-pedition station.)
- QRX – When will you call me again? Or, I will call you again at _____. (Also means standby.)
- QRZ – Who is calling me? Or, you are being called by_____.
- QSB – Are my signals fading? Or, your signals are fading.
- QSK – Can you hear me between your signals? Or, I can hear you between my signals. (A method of CW operation)
- QSL – Can you acknowledge receipt? Or, I am acknowledging receipt. (Also stands for: understood or copied)
- QSO – Can you communicate directly with me? Or, I can communicate with you directly. (A conversation)
- QST – General call preceding a message addressed to all amateurs and ARRL members.
- QSX – Will you listen to me on _____ kHz? Or, I am listening to you on_____ kHz. (Refers to a "split mode" frequency)
- QSY – Shall I move to a different frequency? Or, please change your frequency.
- QTH – What is your location? Or, my location is_____.

The RST[79] (Q) code below is the last of the brevity codes:

Readability
1 – Unreadable.
2 – Barely readable.
3 – Readable with lots of difficulty.
4 – Readable with barely any difficulty.
5 – Perfectly readable.

Signal Strength
1 – Faint signals, barely able to copy.
2 – Very weak signal.
3 – Weak signal.
4 – Fair signal.

5 – Fairly good signal.
6 – Good signal.
7 – Moderately strong signal.
8 – Strong signal.
9 – Extremely strong signal.

Tone
1 – Very rough and broad.
2 – Very harsh and broad.
3 – Rough tone.
4 – Rough note.
5 – Strong ripple.
6 – Trace of ripple.
7 – Near pure tone.
8 – Near perfect tone.
9 – Perfect tone.

Gary L. Drasch

Chapter 13

Before Pressing that PTT

At this point, I hope you are chomping at the bit to either get back on the air or get your license. I'm going to guess you have a working radio and antenna, and maybe you have been doing some listening. Likewise, I have the expectation you hold a valid General Class license or above. Maybe you are not necessarily proficient but still able to pound some brass, copy a few Q-Signals and rattle off some phonetics. You might be that radio amateur who has done his share of operating in the past but has not been on the air for the last thirty or forty years. Your station might fall into the vintage class or maybe you decided this is a sure thing and purchased a shiny new rig. Excellent! You are going to be surprised by how alive and vibrant this remarkable hobby is. If you are a new licensee, WELCOME!

The following is a checklist for you to use before pressing the PTT or footswitch:

1. The first thing you want to do is obtain a copy of the current band plan. It is a chart showing what frequencies and modes are usable with the license you possess. The ARRL offers nice printable charts at http://www.arrl.org/graphical-frequency-allocations. Years ago, this type of list was a chart titled The Amateur Bands and was found in the first chapter of *ARRL handbooks*. Take notice of the current band plan in that there are four additional HF bands available, i.e., 60m, 30m, 17m, and 12m compared to when you likely operated last. Please refer to Chapter 3 for more information.

2. You should know where you can legally operate after reviewing the band plan. I'm going to guess you would start out attempting to make some phone contacts before using CW. The rule of thumb is to use LSB[80] (lower-sideband) when operating 40, 75, and 160-meters. When operating on 60, 20, 17, 15, 12, 10, and 6-meters, you should use USB. Although the VHF and UHF bands are well known for using FM and repeaters, some operators use USB in the allotted portion of those bands too. If you resurrected an AM station, you will find yourself welcomed down on 75 and 160-meters by other AM-ers. You need to look for them though. I expect you know it is not cool to call a SSB station using AM. You should also be aware that 30-meters is reserved for CW and digital modes only. It is also restricted to 200 watts PEP or less. Please refer to Chapter 7 regarding the digital modes.

3. Let us cover a few nuances before flipping the switch. A couple of things changed while I was away from HF for forty-plus years. One of the first is we no longer need to log each-and-every transmission. This changed on June 9, 1983.[81] In fact, you do not need to log any transmissions if you do not want to. The way we interface with each other on the air has changed a little bit too since the 1960s and 70s. Nothing outrageous, but worth noting before you call CQ. CQing has not changed at all, but when an OM or YL responds, they typically do not use your call sign first, followed by "this is" and their call sign. Instead, they only give their call sign (it is assumed they are calling you). At that point, you say their call sign followed by "this is" and your call sign. There is nothing wrong with doing it the proper way— it's just not as common.

4. In the past, if a W3 returned your call, we could count on him being in Pennsylvania, Delaware, Maryland, or DC. Not anymore. In 1978,[82] the FCC dropped the requirement of a

licensee to change their call sign when moving into a different call area. It's now remarkably common to hear W1s, K9s, and AA3s having a QTH in Florida. We now find 6s in the 4s call area, 2s in 7s, and so on. With the change, approximately 88%[83] of the call signs you hear will still reside in the correct numbered area.

5. If you plan to operate CW, which I hope you do, a signal report is no longer sent as 599 but rather 5NN. We spoke about "Cut Numbers" earlier. When sending your name, it is no longer, "Name is Lou" but rather "Name Lou" or "OP Lou." With the QTH, it is "QTH WI" or maybe "QTH Port Washington, WI." There is no "is" between QTH and your city or state. There is much more shorthand taking place than years ago. It is not necessarily a rule or procedure, but rather something which has evolved over time.

6. DX-ing is near and dear to me. You probably figured that out by the amount I wrote on DX-ing,. I covered this earlier but will remind you again. If you are planning to work DX, it is ultra important to etch the following into your brain: not all, but many DX stations, especially if they are rare, operate in a split mode. For example, if you hear a DX station and become elated because you don't hear anyone calling him, STOP! Do not call him—at least not on that frequency. The reason you are not hearing others calling is that the DX-ers are calling him on a different frequency, i.e., the DX station is transmitting on one frequency and he/she is listening on a different one, hence the term split. On SSB, listen 5 to 10-KHz up and you will find the pileup of stations calling him. Depending on the rarity of the DX, the pileup can be anywhere from 5 to 15, even 20-KHz wide. It will sound like a zoo! On CW, it starts at least 1 if not 2-KHz up from where you hear the DX station and might get as wide as 5 to 10-KHz. Take time to listen to the DX. And then, LISTEN again! He should be saying "UP" after his call sign. The reason I said

not to call the DX on his frequency—the one you heard him on—is because the "pileup," i.e., everyone else is listening there. You don't want to interfere! Therefore, wiggle your way into the pileup and transmit on a frequency in the same vicinity as everyone else. You might be in for a big surprise if you were to call him on his own frequency! Please refer to Chapter 11, Beyond Lids.

7. Do not let the bad behavior get to you. As referenced in the Beyond Lids chapter, do not get into a confrontation with the QRM-ers. Do not fuel their fire. Move somewhere else and if worse comes to worst, turn your radio off and come back later. It is not worth letting a few simpletons get to us and spoil our fun. Eventually, they will give up and go away.

By possessing a valid license, you are qualified to get back on the air and start making some contacts. Do not forget you are the captain of your station and ultimately responsible for the correct and legal operation of it. Make sure you are aware of the rules and regulations applying to your license classification. Answers can be found on the ARRL website at http://www.arrl.org/part-97-amateur-radio.

If you do not have a license, now is the time to purchase that study guide, read it, and take the exam. Find a local club and ask for help. The members will be more than happy to assist.

Hope to work you soon!

73 and God bless...

Gary
K9DJT

Acknowledgments

If it had not been for Lyle Ten Pas, WE9R, I would have thought this hobby had long died. It was Lyle who rekindled my interest in ham radio. I am deeply indebted, beyond that of being a good friend, for being my teacher, mentor, and Elmer. Thank you for answering so many of my questions with a question, allowing me to figure things out on my own. No matter if our conversation was ham radio, fishing, hunting, carpentry, dogs, work, or commiseration; we always managed to learn something from each other. If nothing else, the pizza and humor have always been great! Thank you, Lyle.

And if not for Bill Shadid, W9MXQ, and his wife Jean, I would have never considered writing a book. Bill began coaxing me to write after Jean mentioned she enjoyed reading my articles in our local radio club newsletter. Well, here it is! I have always referred to Bill as a boat-anchor historian, and need to thank him for his expertise and assistance with my "Shack Heaters to Now" chapter. He also deserves a big thank you for his general opinion of content and encouragement. Thank you for being a good friend, for the nudge to write, and for all the endless conversations and laughs.

Many anecdotes within involve members of the *Greater Milwaukee DX Association* and the *Ozaukee Radio Club*. All of you have had a positive influence on me in one way or another. I thank all of you for allowing me to share those little stories within the context of this book.

Most importantly, I want to thank my lovely wife Chris for tolerating my obsessive behavior with my various hobbies and projects. Somehow she's managed to put up with it for 50 years. Thank you for your love and for accepting me for who I am. Love you...

Gary L. Drasch

About the Author

Gary Drasch acquired his radio amateur Novice license in 1960 at age 12 and the General class at 13. After graduating from high school, he attended Milwaukee Area Technical College and received a diploma in Electronics, Radio, and Television. He continued his education at MATC by earning a two-year certificate in Radio Communications and an FCC commercial First Class Radiotelephone Operator License. It was considered the pinnacle of licenses at the time as it was required to operate and maintain TV/radio broadcast transmitters. His career began as the owner-operator of a TV, audio, and radio communications business. In 1985, after eighteen years as an entrepreneur, and spending a short stint in the radio pager business, he joined the John Fluke Manufacturing Company as a Sales Engineer. He retired 24 years later at the age of 62 and immediately returned to ham radio. As well as passing the Extra class license exam, he became totally reabsorbed by the hobby. DX-ing, working meteor-scatter, chasing the Fred Fish Memorial Award (FFMA), antenna experimentation, and home-brewing accessories are his passion. Thus far, he has worked 313 countries/entities and 1824 band slots; earned the DX Century Club (DXCC) award in mixed, CW, phone, and digital modes; captured the Worked All Continents (WAC) award, and the Work Antarctica Callsigns Award (WACA). Additional awards include the Worked All States (WAS) Triple Play award and the VHF UHF Century Club (VUCC) award having worked 519 grid squares—the FFMA grid total is 386. Memberships include the American Radio Relay League, the Greater Milwaukee DX Association, Badger Contesters, the Society of Midwest Contesters, and the Straight Key Century Club. He is also a member and Past President of the Ozaukee Radio Club. Gary and his wife Chris have been married for 50 years and live in Port Washington, Wisconsin. They have two adult children, Jennifer and Andrew—and a dog by the name of LuLu.

Bibliography

1. "History of Chess." *Wikipedia*. Wikimedia Foundation, 02 Sept. 2017. Web. 22 Sept. 2017. <https://en.wikipedia.org/wiki/History_of_chess>.

2. SPERONI, JOSEPH P. "Total Amateur Radio Station Licenses (by Month by Class)." *Amateur Radio Station Statistics*. Web. 22 Sept. 2017. <http://ah0a.org/FCC/Licenses.html>.

3. *Spring 1962 Radio Amateur Callbook Magazine*. 1st ed. Vol. 40. Chicago 39: Herbert J. Nelson, 1962. Print.

4. World Amateur Radio Day *International Amateur Radio Union*. Web. 18 April 2021 < https://www.iaru.org/on-the-air/world-amateur-radio-day/ >. Three million radio amateurs.

5. "History of the U.S. Telegraph Industry." *EHnet*. Web. 22 Sept. 2017. <http://eh.net/encyclopedia/history-of-the-u-s-telegraph-industry/>.

6. Ratzlaff, Noji. "CAT - Yaesu." *Ham Radio Glossary*. Web. 22 Sept. 2017. <http://noji.com/hamradio/glossary.php>.

7. Sutcliffe, Gary C. "The Contest Card". *QST Magazine*. Newington: American Radio Relay League, September (1993): 31-32. Print.

8. "Year 1920 - Callbook." *Amateur Radio History*. Web. 22 Sept. 2017. <http://ac6v.com/history.htm>.

9. "The Radio Amateur Callbook." *Callbook - Home*. Web. 22 Sept. 2017. <http://www.callbook.biz/index.html>. CD Available

10. *About by QRZ.COM*. Web. 22 Sept. 2017. <https://www.qrz.com/page/about.html>. Founded in 1992

11. "History of DX-ing." *History of DX-ing | Mile-Hi DX Association*. Web. 22 Sept. 2017. <http://www.mhdxa.com/node/135>.

12. "The History of Yasme." *The Yasme Foundation*. Web. 22 Sept. 2017. <http://www.yasme.org/about-us/>. Danny Weil, VP2VB

13. "DX Century Club." *Wikipedia*. Wikimedia Foundation, 04 July 2017. Web. 23 Sept. 2017. <https://en.wikipedia.org/wiki/DX_Century_Club>.

14. "The Last Man Standing." Personal email exchanges with: Bill Smith, W9VA; John Becker, K9MM; Dave Patton, NN1N; Kermit Carlson, W9XA; June 1-6 (2017).

15. "DXCC Honor Roll." Web. 23 Sept. 2017. <http://www.arrl.org/files/file/DXCC/rule%201_u_v.pdf>.

16. "ARRL." *DXCC Challenge*. Web. 23 Sept. 2017. <http://www.arrl.org/dxcc-challenge>.

17. Not Applicable

18. "Don Miller - W9WNV - Video - OH2BH - Interview." *DX News*. 04 Mar. 2017. Web. 23 Sept. 2017. <https://dxnews.com/w9wnv-oh2bh/>.

19. Panoramic Radio Corp.; New York (NY). "History of the Manufacturer ." *Panoramic Radio Corp.; New York Manufacturer in USA, Model T*. Web. 23 Sept. 2017. <http://www.radiomuseum.org/dsp_hersteller_detail.cfm?company_id=10685>. Marcel Wallace, 1930, Spectrumscope

20. Not Applicable

21. "WARC Bands." *Wikipedia*. Wikimedia Foundation, 02 Sept. 2017. Web. 23 Sept. 2017. <https://en.wikipedia.org/wiki/WARC_bands>.

22. "60-meter Band." *Wikipedia*. Wikimedia Foundation, 02 Sept. 2017. Web. 23 Sept. 2017. <https://en.wikipedia.org/wiki/60-meter_band>.

23. "Frequency Band Chart 8x11." ARRL, 25 Apr. 2017. Web. <www.arrl.org/files/file/Regulatory/Band+Chart/Hambands4-web-color_4-25-17>.

24. "W5YI - What Are Call Sign Groups." *W5YI : Resources for Amateur & Commercial Radio*. Web. 23 Sept. 2017. <http://w5yi.org/page.php?id=281>. Vanity signs - 1978

25. *The Radio Amateur's Handbook*. 39th ed. West Hartford: American Radio Relay League, 1962. Print. Cyclic variations in the ionosphere.

26. Troster, John G., and Robert S. Fabry. ""The NCDXF/IARU International Beacon Network-Part 1"." *QST Magazine* (1998): IARU (International Amateur Radio Union. Web. <http://www.iaru.org/uploads/1/3/0/7/13073366/beacon1.pdf>.

27. Smith, Peter G. "A Short History of the Reverse Beacon." *History of the Reverse Beacon Network - Reverse Beacon Network*. Web. 23 Sept. 2017. <http://www.reversebeacon.net/pages/A+Short+History+12>

28. Andrea, Steve Sant. ""When Should I Operate?"." *QST Magazine*. Newington: American Radio Relay League, July (2010): 68-69. Print. Propagation

29. "List of Solar Cycles." *Wikipedia*. Wikimedia Foundation, 29 Aug. 2017. Web. 23 Sept. 2017. <https://en.wikipedia.org/wiki/List_of_solar_cycles>. Recordkeeping - 1755.

30. Codella, Christopher F. "The Fourth Time's the Charm." *Ham Radio History*. ARRL, 25; August 2013. Web. 23 Sept. 2017. <http://w2pa.net/HRH/the-fourth-times-the-charm/>. Transatlantic communications

31. Codella, Christopher F. "Two Contests." *Ham Radio History*. ARRL, 16 July 2016. Web. 23 Sept. 2017. <http://w2pa.net/HRH/two-contests/>. 1928 International Relay Party

32. "American Morse Code." *Wikipedia*. Wikimedia Foundation, 18 Sept. 2017. Web. 23 Sept. 2017. <https://en.wikipedia.org/wiki/American_Morse_code>.

33. Stenersen, Sigurd. "Just Learn Morse Code." *Koch's Method*. 2005. Web. 24 Sept. 2017. <http://www.justlearnmorsecode.com/koch.html>.

34. Shovkoplyas, Alex. "Morse Runner." *DX Atlas: Amateur Radio Software*. Afreet Software, Inc., 1998. Web. 24 Sept. 2017. <http://www.dxatlas.com/MorseRunner/>.

35. Kolpe, Mathias, DL4MM, and Alessandro Vitiello, IV3XYM. "*The Official RufzXP Homepage by Mathias Kolpe, DL4MM and Alessandro Vitiello, IV3XYM.*" RufzXP-Tancredi, 20 Nov. 2013. Web. 24 Sept. 2017. <http://www.rufzxp.net/>. Morse code call sign training.

36. Wilson, Mark J. *The ARRL Operating Manual for Radio Amateurs*. 9th ed. Newington, CT: American Radio Relay League, 2009. Print. Contesting abbreviations.

37. "QSL Bureaus." *QSL Bureaus - Amateur-radio-wiki*. Web. 24 Sept. 2017. <http://www.amateur-radio-wiki.net/index.php?title=QSL_Bureaus>.

38. "Year 1920 - QSL Bureau Established." *Amateur Radio History*. Web. 24 Sept. 2017. <http://ac6v.com/history.htm>.

39. EQSL.cc. "EQSL.cc Home Page." *EQSL.cc - The Electronic QSL Card Centre*. EQSL.cc, Web. 24 Sept. 2017. <https://www.eqsl.cc/qslcard/About.cfm>. Also, see 10thAnniversary link.

40. Mills, Wayne. "Introducing Logbook of The World." *QST Magazine*. Newington: American Radio Relay League, October (2003): 46-47. Print.

41. Wells, Michael, G7VJR. "The History of Club Log." *G7VJR's Blog*. Web. 24 Sept. 2017. <http://g7vjr.org/2015/09/a-brief-history-of-club-log/>.

42. Koch, Bernd, DF3CB, and Chris, DL5NAM Sauvageot. "Online QSL Request Services (OQRS)." *Oqrs.net : Online QSL Request Services (OQRS)*. Web. 24 Sept. 2017. <http://df3cb.com/oqrs/>.

43. "Radioteletype." *Wikipedia*. Wikimedia Foundation, 15 Sept. 2017. Web. 24 Sept. 2017. <https://en.wikipedia.org/wiki/Radioteletype>.

44. Taylor, Joe K1JT. "WSJT (amateur Radio Software)." *Wikipedia*. Wikimedia Foundation, 26 Aug. 2017. Web. 24 Sept. 2017. <https://en.wikipedia.org/wiki/WSJT_(amateur_radio_software)>.

45. "WSPR (amateur Radio Software)." *Wikipedia*. Wikimedia Foundation, 15 June 2017. Web. 25 Sept. 2017. <https://en.wikipedia.org/wiki/WSPR_(amateur_radio_software)>.

46. Langner, John, WB2OSZ. "Slow Scan TV - It Isn't Expensive Anymore!" *QST Magazine* (Jan.-1993): 20-30. Print.http://www.arrl.org/files/file/Technology/tis/info/pdf/19320.pdf

47. "RMS Express with WINMOR Now Available for Winlink 2000." *ARRL News*, 22 June 2010. Web. 25 Sept. 2017. <http://www.arrl.org/news/rms-express-with-winmor-now-available-for-winlink-2000>.

48. "High-altitude Balloon." *Wikipedia*. Wikimedia Foundation, 09 Aug. 2017. Web. 25 Sept. 2017. <https://en.wikipedia.org/wiki/High-altitude_balloon>. Amateur radio high-altitude ballooning.

49. "Automatic Packet Reporting System." *Wikipedia*. Wikimedia Foundation, 17 Sept. 2017. Web. 25 Sept. 2017. <https://en.wikipedia.org/wiki/Automatic_Packet_Reporting_System>. Bob Bruninga, WB4APR, 1980.

50. Ford, Steve. *The ARRL Satellite Handbook*. First ed. Newington, CT: American Radio Relay League, 2010. Print.

51. Lepre, Lyn. "Ham Radios in Space." *NASA*. NASA, Web. 25 Sept. 2017. <https://science.nasa.gov/science-news/science-at-nasa/2000/ast21aug_1/>. 1997-Ham radio equipment; part of payload.

52. "Amateur Radio Licensing in the United States." *Wikipedia*. Wikimedia Foundation, 21 Aug. 2017. Web. 25 Sept. 2017. <https://en.wikipedia.org/wiki/Amateur_radio_licensing_in_the_United_States>. Volunteer examiners.

53. *The Radio Amateur's License Manual*. 54th ed. Newington, CT: American Radio Relay League, 1965. Print.

54. "The ARRL VEC: More than Just Amateur Radio Exams." *ARRL*, Web. 25 Sept. 2017. <http://www.arrl.org/news/the-arrl-vec-more-than-just-amateur-radio-exams>. 1984-Volunteer examiners.

55. "Amateur Radio Licensing in the United States." *Wikipedia*. Wikimedia Foundation, 21 Aug. 2017. Web. 25 Sept. 2017. <https://en.wikipedia.org/wiki/Amateur_radio_licensing_in_the_United_States>. Restructuring in 2000, End of Morse code requirement, Current License Classes.

56. "Field Day (Amateur Radio)." *Wikipedia*. Wikimedia Foundation, 10 June 2017. Web. 25 Sept. 2017. <https://en.wikipedia.org/wiki/Field_Day_(amateur_radio)>. Started in 1933.

57. "New to Field Day? START HERE!" *Field Day*. ARRL, Web. 25 Sept. 2017. <http://www.arrl.org/field-day>. 40,000 participants.

58. "Amateur Radio Emergency Service." *Wikipedia*. Wikimedia Foundation, 02 May 2017. Web. 25 Sept. 2017. <https://en.wikipedia.org/wiki/Amateur_Radio_Emergency_Service#Historical_operations>. Formerly the AREC, started in the 1930s.

59. Maxwell, Jim, W6CF. "Amateur Radio: 100 Years of Discovery." *QST Magazine*. Newington: American Radio Relay League, January (2000): 28-34. Print. The '70s—Repeaters and Packets.

60. Taylor, Jonathan, K1RFD. *ARRL's VoIP: Internet Linking for Radio Amateurs*. First ed. Newington, CT: American Radio Relay League, 2004. Print.

61. "D-STAR." *Wikipedia*. Wikimedia Foundation, 22 Sept. 2017. Web. 25 Sept. 2017. <https://en.wikipedia.org/wiki/D-STAR>.

62. Balister, Roger, G3KMA. "Geoff Watts, Founder of IOTA." Web. 25 Sept. 2017. <http://www.dokufunk.org/upload/watts_geoff.pdf>.

63. "Introduction - a Brief History of SOTA." *About SOTA*. Summits On The Air, Web. 25 Sept. 2017. <http://www.sota.org.uk/About>.

64. Varetto, By: Gianni, By: Betty Sciolla, and KK3Q By: Floyd Larck. "W.A.P. – W.A.P. Worldwide Antarctic Program." *WAP*. Web. 25 Sept. 2017. <http://www.waponline.it/>. History.

65. "Maidenhead Locator Squares." *Historical Background*. QSL.NET, Web. 25 Sept. 2017. <http://www.qsl.net/ei7gl/locsqr.htm>.

66. "VHF/UHF Century Club Award Rules." ARRL, Web. <https://www.arrl.org/files/file/Awards%20Application%20Forms/VUC CRULE1a.pdf>.

67. Brickey, Don, W7OK. "County Hunting History." ICHN Annual Covention, Peoria, IL - 1972, Web. 27 Sept. 2017. <http://www.charchive.com/topics/stuff/peoria72.pdf>. pdf File.

68. "The Award Numbers." *The Numbers*. County Hunter Dot Com, 20 June 2017. Web. 26 Sept. 2017. <http://www.countyhunter.com/num_frame.htm>.

69. "Yaesu (brand)." *Wikipedia*. Wikimedia Foundation, 30 Aug. 2017. Web. 26 Sept. 2017. <https://en.wikipedia.org/wiki/Yaesu_(brand)>. Arrived in California 1965.

70. NoobowSystems. "SideBand Engineers - The Company." NoobowSystems Lab, Web. 26 Sept. 2017. <http://www.noobowsystems.org/restorations/sb-33/sbe-company-e.html>.

71. "iCom Global-History." *History | Company Information*. iCom, Web. 26 Sept. 2017. <https://www.icom.co.jp/world/company_profile/history/>.

72. Vigil, Sam, WA6NGH. "Interview with 'Mr. ICOM' Tokuzo Inoue, by CQ Amateur Radio Magazine." *Interview with 'Mr. ICOM' Tokuzo Inoue, by CQ Amateur Radio Magazine | Icom Inc*. CQ Magazine, Web. 26 Sept. 2017. <http://www.icom.co.jp/world/news/004/>.

73. "After the dust settled in 1984," Personal, one-on-one, interviews regarding equipment history and the evolution into present-day ham radio stations: William Shadid, W9MXQ, April 20-June 1 (2017).

74. "Q Code." *Wikipedia*. Wikimedia Foundation, 14 Sept. 2017. Web. 26 Sept. 2017. <https://en.wikipedia.org/wiki/Q_code>.

75. "The RST (Readability-Strength-Tone) System – 59!" QRZ Now - Amateur Radio News, 30 June 2015. Web. 26 Sept. 2017. <http://qrznow.com/rst-readability-strength-tone-system-59/>.

76. Fair, Walt, W5ALT. "Ham Radio Slang." Walt Fair, W5ALT, Web. 26 Sept. 2017. <http://www.comportco.com/~w5alt/index.php?pg=5>.

77. Fields, Ronald, W5WWW. "MORSE CODE (CW) ABBREVIATIONS." *Morse Code (CW) Abbreviations*. Web. 26 Sept. 2017. <http://www.qsl.net/w5www/abbr.html>.

78. Butler, Donald, N4UJW. "Q Signals, Prosigns and Abbreviations For The Ham Radio Operator." *Q Signals, Prosigns, and Abbreviations For The Ham Radio Operator*. Hamuniverse, Web. 26 Sept. 2017. <http://www.hamuniverse.com/qsignals.html>.

79. "Quick Reference Operating Aids, The RST System." ARRL, Web. 26 Sept. 2017. <http://www.arrl.org/quick-reference-operating-aids>.

80. Silver, Ward, N0AX. "How to Listen to Ham Radio on Single Sideband." *Dummies*. Web. 26 Sept. 2017. <http://www.dummies.com/programming/ham-radio/how-to-listen-to-ham-radio-on-single-sideband/>. Upper or Lower sideband?

81. Smith, Carol, AJ2I. "Most Logging Eliminated." *QST Magazine*. Newington: American Radio Relay League, August (1983): 56-57. Print.

82. "History of the Ham Radio Call sign." e*Ham.net - Amateur Radio (Ham Radio) Community*. EHam, Web. 26 Sept. 2017. <http://www.eham.net/articles/38849>. February 23, 1978-Amateurs are no longer required to change their call sign when moving to a new district.

83. "Amateur Radio Licensing in the United States." *Wikipedia*. Wikimedia Foundation, 21 Aug. 2017. Web. 26 Sept. 2017. <https://en.wikipedia.org/wiki/Amateur_radio_licensing_in_the_United_States>. Subject: Call signs.

www.ingramcontent.com/pod-product-compliance
Lightning Source LLC
Chambersburg PA
CBHW020659220526
45464CB00001B/498